Amber
Jewelry, Art & Science

Nancy P. S. Hopp, M.Ed., M.V.

Schiffer Publishing Ltd

4880 Lower Valley Road Atglen, Pennsylvania 19310

Dedication

This book is dedicated to my father,
James B. Purcell, an avid writer,
who encouraged me to follow in his footsteps.
His love of writing is now mine.

covers designed by Bruce Waters
Type set in Goudy Oldstyle

ISBN:978-0-7643-3168-8

Printed in China

Schiffer Books are available at special discounts for bulk purchases for sales promotions or premiums. Special editions, including personalized covers, corporate imprints, and excerpts can be created in large quantities for special needs. For more information contact the publisher:

Published by Schiffer Publishing Ltd.
4880 Lower Valley Road
Atglen, PA 19310
Phone: (610) 593-1777; Fax: (610) 593-2002
E-mail: Info@schifferbooks.com

For the largest selection of fine reference books on this and related subjects, please visit our web site at
www.schifferbooks.com
We are always looking for people to write books on new and related subjects. If you have an idea for a book please contact us at the above address.

This book may be purchased from the publisher.
Include $5.00 for shipping.
Please try your bookstore first.
You may write for a free catalog.

In Europe, Schiffer books are distributed by
Bushwood Books
6 Marksbury Ave.
Kew Gardens
Surrey TW9 4JF England
Phone: 44 (0) 20 8392-8585; Fax: 44 (0) 20 8392-9876
E-mail: info@bushwoodbooks.co.uk
Website: www.bushwoodbooks.co.uk
Free postage in the U.K., Europe; air mail at cost.

Contents

Acknowledgments

My thanks go to friends and family members who have encouraged me in this endeavor, read the manuscript, shared amber they owned, offered invaluable suggestions, and helped to photograph this amazing material.

Special thanks go to my friend, Patty Farrell, MFA, for her drawings; to my son David Shively for his assistance with maps and charts; to my son James Shively for going on amber excursions with me; and to my husband for his encouragement and understanding.

A jewelry chest overflowing with amber delights standing next to a sundial portrays the timeless beauty of this amazing stone. *Author's collection.*

Preface

My interest in amber began in childhood during a visit to the Sponge Docks at Tarpon Springs, Florida, where exhibits and shops attract tourists. My mother found an amber necklace and earrings set she liked, of plain yellow-orange color. The shop owner shared the virtues of the pieces, and I began to take an interest. He rubbed the beads on his wool shirt and held them up to my mother's hair. Her hair stood on end; it was attracted toward the beads. Next, he took a sheet of paper and tore it into small pieces. Again he rubbed the beads and placed them next to the paper. Amazingly, the paper moved toward the beads. We had been shown one of the secrets of this amazing stone. My mother purchased the "mutton fat amber" necklace and earrings and enjoyed wearing them the rest of her life. My interest in the world of amber began, as it had for prehistoric man centuries before, with an unexpected discovery.

As an adult, I took a silversmithing course. Working with silver and gold was exciting, but working with stones was even more so. I was fascinating to discover that amber was one of the more difficult stones to set. Its low melting point means it cannot be set using a torch, and it scratches easily. "Cherry amber" became a special interest of mine.

Years later, at a juried art show, I met an artist who created beautiful jewelry. I attempted to purchase a piece for my daughter and was informed that he needed to do muscle reflex testing on her before selling me the piece. I had no idea what he was talking about, but brought my daughter to the show the next day. He used kinesiology to determine which jewelry was good for her. Again, a great salesman had sparked my interest. We purchased two pendants and I began to study the power of stones and the impact that color has on our lives. Psychology, metaphysical, and gemology books became part of my library. The unique properties of amber and other stones stimulated my interest in a new way.

The University of South Florida offered a Gemology Certificate and a Master Valuer Program in conjunction with the California Institute of Jewelry Training. My love of education and interest in jewelry made the classes fascinating. I took the classes needed for gemology certification and enrolled in the Master Valuer Program. The instructor, Dr. Eva Ananiewicz, explained how to look at, test, research, and appraise jewelry. I wrote the required term paper on amber and my interest in the stone was rekindled.

After more years of research, I can't believe all the exciting and interesting discoveries I have made about this unique treasure. This book is not written as a technical report, but as a story relating amber's unique properties, fall in and out of favor, and history. As I develop a more serious approach, I find amber to be a wonderful compulsion, filled with interesting people and amazing facts.

Esther Purcell was fascinated by this "mutton fat" amber necklace and earrings. *Author's collection.*

Cherry amber necklaces were a favorite at the antique malls in Tarpon Springs. This faceted "cherry amber" is really Bakelite. *Author's collection.*

IDENTIFICATION OF AMBER

What is Amber?

The mineralogical name of amber is succinite. It is one of a few stones that owes it origin to organic life and is considered a Class VIII Mineral, and therefore a stone. (Schumann, 1993, p. 32-33) Not only was this stone a creation of organic matter, but it is also the only stone that contains organic inclusions. Pieces of plants and animals living over fifty-million years ago were trapped in the sticky resin and give us a glimpse of life during that long lost time. (Schumann, 1997, p. 50)

Amber is an organic, gummy resin, secreted from special conifer and other trees millions of years ago. As this ancient resin ages and is exposed to the atmosphere, it begins to darken and become crusty. After millions of year's exposure, it breaks down into tiny shards and fragments. The oldest known amber, from the Isle of Wight, UK, is about 120 million years old. These pieces are very small and weigh only a few grams. Baltic amber is about 40 million years old and can still be found in large pieces. Copal from Africa, Columbia, and Madagascar is much younger; only about one to three million years and is still found in larger chunks. Amber that has remained buried and has not been subjected to weathering from the atmosphere will not break down as readily. (www.ganoskin.com/orchid/archive/199702/msg00073.htm)

Now, let's look at a more technical description of this amazing stone.

Its mineral name is succinite. Amber, like other resins, consists essentially of carbon, hydrogen, and oxygen. These elements are combined together in a somewhat variable proportion, but on the average the material contains 79% carbon, 10.5 % oxygen, and 10.5 % hydrogen. This composition is represented by the chemical formula $C_{10}H_{16}O$. A small amount of sulphur is sometimes present. When burned, a little inorganic material remains behind as ash. Pure amber contains about 1/5 of a percent ash, but when inclusions of foreign substances are present this percentage rises considerably. (Bauer, 1968, 535)

This large 6.3 pound block of butterscotch amber came from the Jhonson's mine in El Valle. It clearly shows separations in the deposit of amber material. *Courtesy Patricio Johnson.*

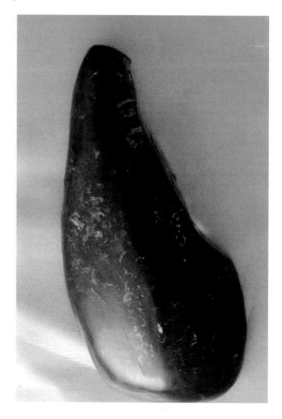

The smaller piece of amber from the La Cumbre region shows color changes that indicate inclusions. Courtesy *Amber Castino.*

Amber is completely insoluble in water, and is only slightly affected after lengthy contact with alcohol, sulphuric ether, acetic ether, and other solvents. This shows an important distinction between true amber and the many similar resins, which are often substituted for it. These resins are quickly attacked by alcohol and the other solvents mentioned above. (Bauer1968,535)

In concentrated sulphuric acid, finely powdered amber is perfectly soluble even when the acid is cold. It is completely decomposed by boiling nitric acid.

When heated, amber gives off a pleasant piney odor. This also occurs when it is rubbed briskly.

It softens at 150 degrees C. and melts between 280 and 290 degrees C.

The Baltic specimen shows little change from the acetone testing.

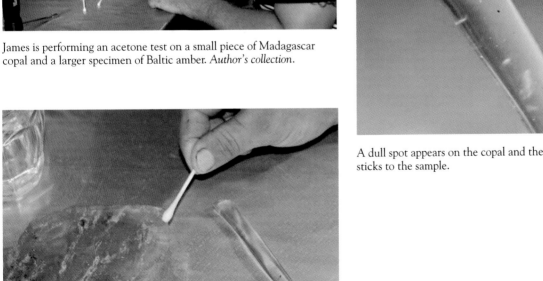

James is performing an acetone test on a small piece of Madagascar copal and a larger specimen of Baltic amber. *Author's collection.*

A dull spot appears on the copal and the q-tip initially sticks to the sample.

A drop of acetone is applied to each specimen and about a minute lapses before checking the samples.

Amber was formed as a viscous liquid that eventually solidified and is therefore an amorphous solid that shows no signs of crystalline structure. It can show flow lines.

Amber was formed as a viscous liquid that was secreted by trees at different intervals. It eventually hardened to create a lump of amber. The yellow amber specimen from La Cumbre shows strong color separation. *Courtesy Patricio Jhonson.*

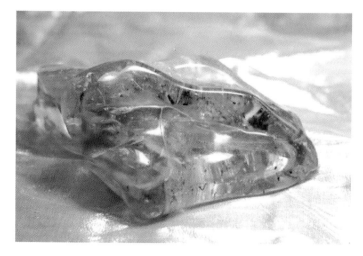

The honey tone specimen shows several amber flows and numerous inclusions. *Courtesy private collector.*

Amber formed in irregular rounded nodules in the form of rods, drops, and flattened plates, but rarely in flat or plane surfaces.

Amber and copal are found in many shapes and sizes. This Madagascar copal is formed in stalactites and rods. *Author's collection*.

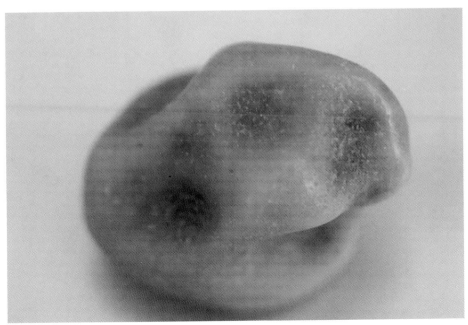

This cloudy piece was formed from a series of drops that came together to form the interesting glob seen here. *Courtesy private collector*.

9

The most common shapes of natural amber are drops and stalactites that dripped from the trees as part of the normal process of resin production. During more abundant resin flows, streamlets and large lumps were produced. Sometimes the amber would be deposited in crevices and voids in the trunks where it would form flat plates. These plate-like forms would be the most pure as inclusions would not settle into the resin. (Rice, 1993,18)

There is no cleavage and a fracture in amber will be conchoidal with a shiny surface that seems oily in appearance.

A specimen in the Amber Mundo Museum resembles bark on a tree. A slice of the material displays the clear amber found under the skin. *Courtesy Amber Mundo Museum.*

When amber is broken a conchoidal fracture appears shiny or waxy. *Courtesy Amber Castino.*

The specific gravity of amber ranges from 1.05 to 1.20 and is only slightly heavier than water. It has a similar S.G. to sea water and will float near the surface in saturated salt water.

The most transparent amber will be denser and have a slightly higher S.G. Amber that is opaque shows turbidity, the presence of air bubbles, and will have the lower S.G. due to the captured air.

Amber is a fairly soft stone with a hardness of about 2 to 2 ½ on the Mohs scale and can be scratched with a finger-nail. Burmese amber or amber from Myanmar is the hardest, a 3 on the Mohs scale. Dominican amber is the softest at 1 to 2. Baltic amber is usually a 2 to 2 ½.

Even though amber is soft, it is a tough stone and thus is useful as a carving material. Its softness and toughness also makes it useful as a bead material because it is easy to drill. Amber is not readily cut as it tends to turn into powder when cutting is attempted. Dominican amber tends to be more brittle and breaks more easily than other ambers.

The specific gravity of amber can be checked using a Universal Specific Gravity Test Kit. The specimen is weighed in the air and in water and then the SG is calculated.

This interesting amber from the Dominican Republic is filled with large and minute bubbles as well as inclusions. *Courtesy Amber Art, D.R.*

The specific gravity of amber is about the same as salt water and amber will float near the surface of a saline solution.

Beautiful bubbles and a termite colony fill this specimen from the El Valle region. *Courtesy Patricio Jhonson.*

11

The refractive index of amber is about 1.54. Most transparent amber is singly refractive because it is an amorphous material. However, if there are inclusions, there might be some double refraction near the inclusion.

Under magnifications, amber is rarely clean. It shows the inclusions and uneven flow that it developed in its creation.

When strike testing amber, it gives a white streak.

It cannot be evaluated using an absorption spectrum.

Michelle is checking the refractive index of the pendant to see if it is amber. Other look-alike materials will have different refractive indices.

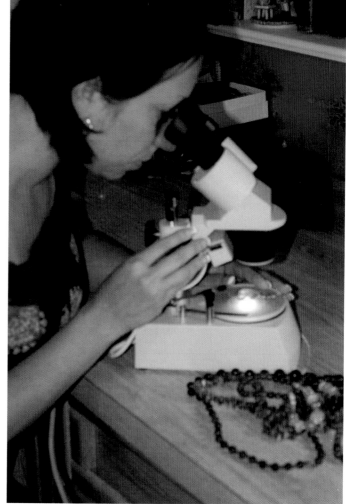

If a microscope is handy, it can be used to get a closer look at the sample. Natural amber will not normally be even in color and should have some bubbles or inclusions. Michelle is looking at the same piece to determine if it is natural amber.

Amber will fluoresce when placed under UV light for a few minutes. The most common colors seen are a bluish-white to yellow-green. Occasionally a piece will fluoresce green, orange, or white. Specimens with higher sulphur content tend to fluoresce more. Fluorescence occurs in only some specimens of amber.

Amber under flourescent light. The amber appears to glow

Most amber will show florescence under ultraviolet, black light. Amber from most locations retains its natural coloring in daylight. The same piece of Dominican amber is photographed in incandescent light, seen here, as well as fluorescent and ultraviolet light.

A color change from yellow to brown in transmitted light to a dark bluish to greenish color in reflected light can occur in some amber. This is interesting to see, but has been considered a distraction in jewelry and is undesirable.

Some Dominican amber is unique in its fluorescence. It fluoresces in daylight and Dominican blue amber makes beautiful jewelry and collectibles.

A Dominican amber necklace shows amber coloration in normal household lighting.

Under ultraviolet lighting, the same blue amber necklace shows an intense bluish glow and milky coloration..

Under ultraviolet lighting ,but close to fluorescent lighting, blue Dominican amber appears blue with milky areas..

One very unique characteristic of amber is that when it is rubbed with a wool or flannel cloth, it becomes strongly charged with negative electricity and attracts scraps of paper, hair, straw, feathers, and other light scraps of similar materials.

Since it melts at a fairly low temperature, it is a bad conductor of heat.

Cold does not penetrate the stone and it usually feels warm to the touch. This is again different from most minerals that are cold to the touch.

Kelly is intrigued that static electricity caused by rubbing a piece of amber on wool can cause a feather to stick to the amber.

This is a detail of the feather clinging to the amber. After a few minutes, the amber lost the charge and the feather fell off.

Most amber takes on a good polish making it desirable for jewelry, but some pieces are naturally dull and will not take a polish. These pieces can be utilized in other ways.

An ivory bird perched on an amber post or stump shows how amber with a dull finish will not overpower the main element, the bird. *Courtesy Frank and Barbara Charbonneau, The Collector's Corner, Nobelton, FL.*

Some amber is naturally dull and does not take a good polish. The cognac and honey amber necklaces in the photo show their polish, but the white amber necklaces have a dull appearance. *Courtesy John McLeod.*

The transparency of amber makes it desirable for jewelry, but some amber is completely opaque and can be used to make unique pieces. In some samples, you might find both characteristics as well as every level between blending into each other. Carvers utilized this characteristic of transparency and opacity to showcase the carver's talents and the beauty of the stone. They situate the object to be carved at just the right place to show subtle color and opacity where it is most needed.

The most common amber in the market place today is Baltic amber, and it is fairly uniform in appearance but varies in color. This amber ranges in color from the palest yellow to dark yellow and even brown. Amber from other locales can show different coloration. Specimens that have gone through surface alterations are sometimes red in color. Amber that has been compressed may appear white in color. Black amber, really very dark shades of other colors, has also been found. Green amber, and red amber, as well as blue and violet, were considered rare, but more of these colors are being discovered and entering the marketplace. Few minerals show this diversity of color.

An artist with an eye for detail studied this monkey and was able to create a beautiful amber sculpture. *Courtesy Amber Mundo Museum.*

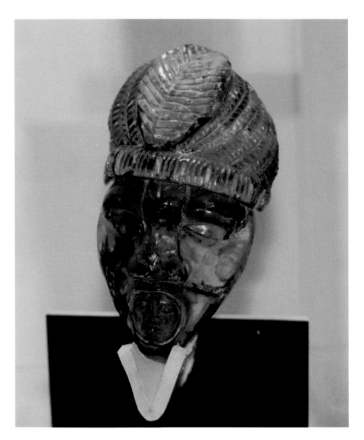

Artists who understand the unique properties of amber create wonderful works of art. This ancient 3 inch tall Taino mask carved in amber is an excellent representation of the art. *Courtesy Amber Mundo Museum.*

17

Artists in the Dominican Republic are still carving amber in the Taino style. This 2-inch tall statue shows the carver's talent. *Author's collection.*

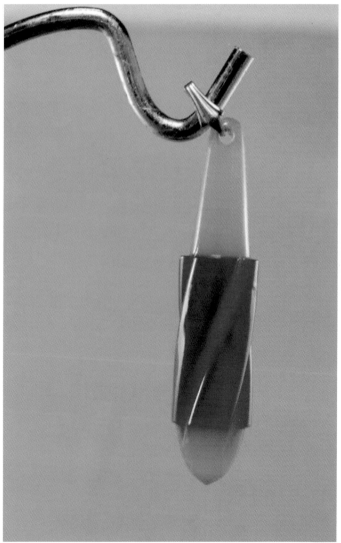

Two tones of amber were effectively used to create this pendant. *Author's collection.*

One other color characteristic of amber is its color change. A form of chemical alteration takes place internally within the amber and pale amber darkens over time. Specimens that were yellow become red or brownish-red. Clear amber becomes slightly darker and cracks begin to develop in the stone. However, amber that has been retained under the seas does not appear to suffer these changes. The color changes are now believed to be attributable to atmospheric weathering. Most minerals do not undergo this color transformation or physical change and this again showcases amber as a unique stone.

Amber that has been subjected to weathering can have this cracked appearance. *Author's collection.*

The Amber Family

The color range in amber is extremely varied. Specimens range from near white through all shades of yellow, into brown and red, as well as blue, green, violet and black. The reasons for the colors are not fully understood, but there are theories to explain them. Some scientists believe that all amber was not created by conifers and that the type of tree predisposed the color of the amber. Scientists at the Polish Museum of Science believe that some reddish amber may have originated from deciduous trees like cherry and plum. (www.amberjewelry.com/SearchResults.asp?Cat=89) Dominican amber with a reddish tint is thought to be related to legumes that may have been near the resin after it was deposited. Volcanic ash is also believed to be a colorant in some amber.

The most typical amber utilized in jewelry is Baltic amber. It is fairly uniform in color and ranges from pale yellow into oranges, reds, and browns. There are other varieties of amber, rarer in the jewelry market, which we should consider.

The following descriptions of the types of amber are derived from Max Bauer's commentaries in *Precious Stones*, Volume II, pp. 537-538).

The amber most people are familiar with is medium brown in color. This beautiful amber shows areas of orange and yellow. *Courtesy Charles Albert jewelry.*

Most people believe amber to be amber in color. Not all varieties of amber are that honey gold color. Several varieties of amber are recognized, the main distinction between them being based principally on the combination of different shades of color with different degrees of transparency. These different ambers also differ in the capacity to acquire a polish. Some amber is more suitable for ornamental purposes than others, and are therefore of greater commercial value.

Transparent amber is considered in the trade to be clear. Shelly amber is transparent amber that is nearly always clear. It is never cloudy throughout and only rarely is cloudiness found at all and it will be in alternate layers. Massive amber, on the other hand, is nearly always more or less cloudy. Perfectly transparent specimens of massive amber are very rare, though they are found more frequent than cloudy specimens of shelly amber. A clear massive variety occurs in masses ranging in colour from almost completely colorless to dark reddish-yellow. Water-clear amber is very rare. This amber is described as yellow-clear. The reddish-yellow or red-clear specimens are found more frequently in nature.

Several varieties of *cloudy* amber are recognized in the lapidary field. They are named: flohmig, bastard, semibastard, osseous, and frothy amber. Some specimens possess characters located somewhere between clear and cloudy. These varieties are distinguished by compound terms of a descriptive kind, such as, clear-flohmig, flohmig-clear, flohmig- bastard, osseous bastard, etc.

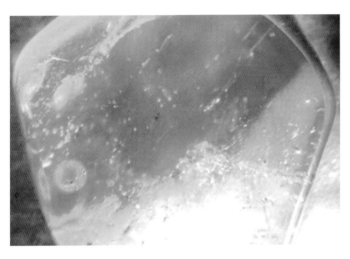

Shelly amber is normally clear, transparent amber, but a detail shows a much more interesting picture. *Author's collection.*

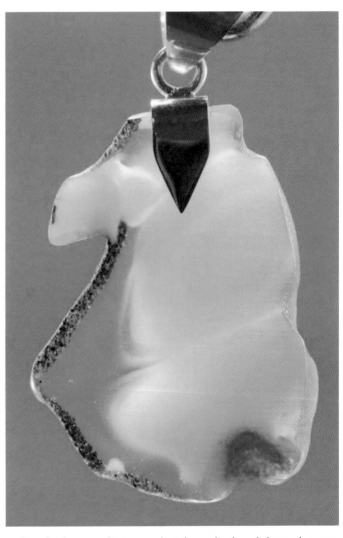

A slice of amber turned into a pendant shows cloudy and clear amber combined with a touch of the natural skin. *Courtesy The Sterling Rose jewelry.*

Flohmig amber is slightly turbid, containing very small air bubbles. It has the appearance of having been clouded by a fine dust and like the clear varieties it is capable of accepting a fine polish. The term 'flohmig' is derived from the east Prussian word 'Flohmfett', which signifies the semi-transparent yellowish fat of the goose or duck. This variety of amber is supposed to resemble goose fat in appearance.

Flohmig amber appears to be clouded with a fine dust. This specimen has the dull luster usually associated with this variety. *Courtesy The Sterling Rose jewelry.*

This red variety is also clouded. When set with a gold druzy and blue Topaz, it makes an interesting pendant. Courtesy The Sterling Rose jewelry.

Bastard amber is considered more turbid. It is still susceptible to a good polish. Various terms are used to describe the turbidity. Material which is cloudy through-out is termed *bastard proper*. Amber in which cloudy portions are dotted about in a clear ground-mass is known as *clouded-bastard*.

An unusual name is given to some amber. The rare piece of amber in this pendant is known as Bastard amber. It shows swirls of white, green, and blue. *Author's collection.*

Another piece of bastard amber has white and blue patches in a shelly, or clear amber background. *Courtesy The Sterling Rose jewelry.*

Color distinctions are also recognized in this variety. Pure white to grayish-yellow shades of bastard-amber are described as *pearl-colored*. The material with paler tones is known in the trades as "blue amber", not to be confused with the rare amber which is actually blue in appearance.

Sometimes you can find the rare Baltic blue amber. One bead on this Lithuanian necklace reflects the unusual blue glow in natural light. *Author's collection.*

Yellow and brownish-yellow bastard amber is called *kumst'-coloured*. The name comes from the East Prussian name Kumst for cabbage (Sauerkraut). The yellow is described as pale and the brownish as dark kumst. Semi-*bastard* amber is somewhere between bastard and osseous amber, combining the appearance of the latter with the capacity for receiving a polish of the former.

Kumst amber is rarely found in jewelry as most Americans prefer the clear amber. This handmade necklace has a beautifully swirled, yellow, kumst amber cabochon. *Courtesy The Sterling Rose jewelry.*

Osseous amber, *bone amber*, is opaque, softer than the varieties described above. It is inferior to them in the ability to receive a polish. Its color ranges from white to brown, and as the name implies, it has the general appearance of bone or ivory.

By combining many of the characters of the different varieties described above, there arise several different colour-varieties of amber, which are classified into two groups under the descriptions variegated osseous clear and variegated osseous bastard.

"*Frothy* amber is opaque amber. It is very soft, and incapable of receiving a polish. It often is found with large numbers of iron pyrite included in the stone.

Bone amber, Osseous amber, was named because of its resemblance to ivory or bone. This small pendant has a touch of clear amber to add a touch of luster. *Courtesy The Sterling Rose jewelry.*

Most bone amber does not have the luster. This exquisite arm band shows the classic look of bone. *Courtesy John McLeod.*

In describing the types of amber, turbidity was discussed. In the past turbidity was believed to be caused by water inclusions in the resin. It is now theorized that the turbidity is actually small bubbles trapped in the matrix. The bubbles are difficult to see with the naked eye, but can be seen under magnification. The size and frequency of the bubbles help determine the type of amber.

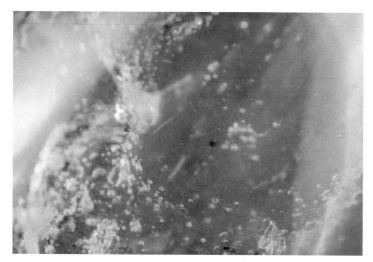

A detail of honey colored amber allows you to see the bubbles found in amber that help to make it so lightweight.

Turbidity is the focal point of this large pendant. A mass of bubbles gives it a unique look. *Courtesy John Mc Leod.*

Turbidity also is a causal factor in the rarer green and blue ambers. *Green* amber ranges in color from pale green to a greenish-black. Some green amber is olive green and other is the apple-green found in chrysoprase. This beautiful color sometimes has clouds of white. Most green amber seen on the market today is yellow amber which has had its back either burned or painted to bring out the green coloration.

The Baltic *blue* amber can be steel blue, azure blue, or sky blue in color. The bubbles in these ambers are one of the reasons for this beautiful and rare color.

True, untreated green amber is rare. Most green amber has been treated to bring out this unusual coloration. *Author's collection.*

The back of this pendant shows how the back was blackened to create the color. Other pieces have the back burnt to create the same effect. *Author's collection.*

The blue amber from the Dominican Republic appears most blue when it has a layer of its crust still intact. As the crust is removed, the vibrancy of the color diminished. Many collectors of this amber only allow a small window to be polished into the stone showing the beautiful coloration. The rest of the piece retains its crust.

Dominican blue amber has a blue coloration in sunlight. Some of the blue shown here also has a reddish coloration. *Author's collection.*

When looking for amber in the market today, you will come across various colors and matching descriptions used to increase your interest in purchasing each color. All the colors are very beautiful and would make a great addition to any amber collection.

Cherry amber is a term used to describe ambers with red coloration. Much "amber" known as cherry is really Bakelite. This 32-inch "cherry amber" necklace is uniform in color and has no inclusions. The piece is Bakelite. *Author's collection.*

Faceted "Cherry amber" is often found in antique stores and flea markets. It is beautiful, collectible, expensive, but most is bakelite. Author's collection.

Some amber turns darker and exhibits a red coloration as it ages. The cloudy amber represented here and strung with 14k gold beads is again bakelite which gives off an unpleasant formaldehyde scent when warmed. *Author's collection.*

Cherry amber from Chiapas Mexico is real, untreated amber. The color is uneven and splotchy. The large pendant was created by *Charles Albert* jewelry. *Author's collection.*

This tiny cherry amber pendant with a sun spangle has been dyed to give it the cherry color. Amber has been dyed for centuries, but real cherry is much more valuable than dyed. *Courtesy The Sterling Rose jewelry.*.

In Poland, there are about 200 folk names given to amber and some eighty variety names. (www.emporia.edu/earthsci/amber/types.htm) I have included some of the more popular colors as seen above and below:

A lot of Baltic amber is heat treated to clear cloudy specimens. This cognac pendant shows the sun spangles or fish scales produced by the treatment. Americans love this look. *Courtesy John McLeod.*

Cognac amber is very popular. The color resembles the richness of cognac. To make the best use of a piece of amber, most jewelry in made free form to contour to the specimen. The triangular form is a great focal point. *Courtesy John McLeod.*

This piece of cognac amber was polished and shaped to form a large drop. A few spangles sparkle to give it life. *Courtesy John McLeod.*

Honey amber is slightly lighter colored than cognac amber. This dark honey piece is framed with a modern sea creature. *Courtesy John McLeod.*

Butterscotch amber is opaque, slightly dark amber with splotches of lighter amber. The amber in this piece goes well with the oxidized sterling. *Courtesy John McLeod.*

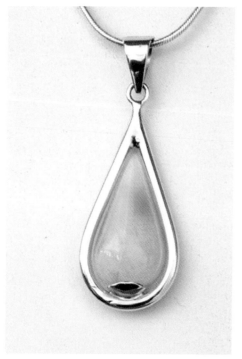

Butterscotch does equally well in a modern design. *Courtesy John McLeod.*

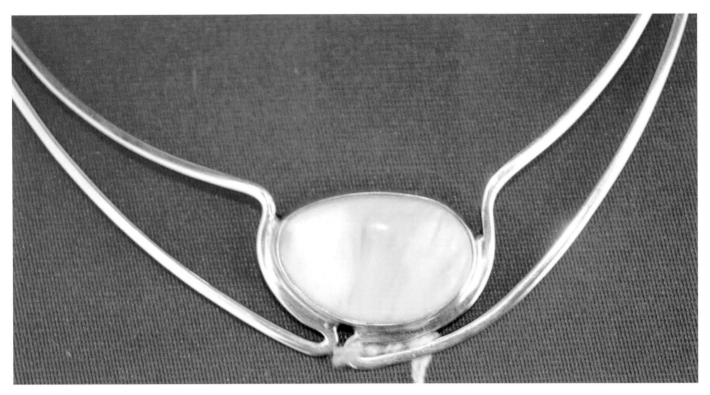

This stunning neck piece has a butter amber focal point. *Courtesy John McLeod.*

Butter amber is lighter than butterscotch. The simplistic design is perfect for this butter cabochon. *Courtesy john McLeod.*

Figure10-31 The Latvian necklace shows a variety of natural colors. A yellow amber bead is flanked with cognac beads. The bead to the right has been heat treated and shows spangles. All the others are natural colors. *Author's collection.*

This large amber pendant could easily be mistaken for bone or another stone. However it is lightweight, osseous amber. *Author's collection.*

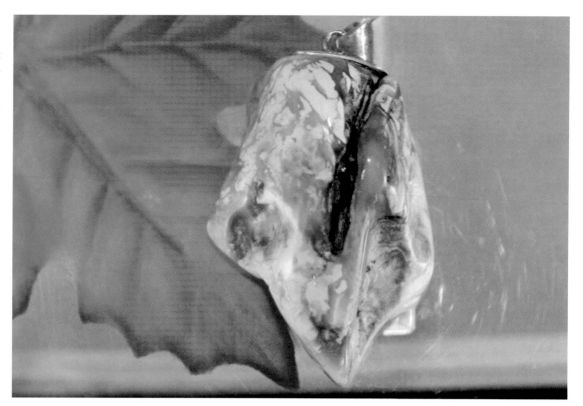

Orange amber glows with the warmth of autumn in this Dominican amber bracelet. The faceted beads make the piece very special as the carving is very time- consuming and difficult. *Author's collection.*

Green or moss amber look great on this vignette pendant. A trip to the winery could be in order. *Courtesy John McLeod.*

Even though the word amber means brownish yellow, amber comes in many colors. In this artist rendering of the Necklace of Galatea, the diversity of amber's colors was known and utilized as early as the late 1800s.

Sea Stone, Scoop Stone, and Pit Amber

Another way of classifying amber is to look at where the amber was collected. Sea stone and scoop stone are ambers found in or near the sea. Scoop stone refers to amber collected or scooped from seaweed; sea stone, or sea amber, has washed up on the beach. Divers would go under the sea and retrieve amber from the seabed as well as scoop it from the surface. They utilized special tools to retrieve the specimens. This sea amber has usually been polished by the action of the waves and sand. (ww.amberjewelry.com/SearchResults.asp?Cat=89)

Pit amber is mined from the blue earth in which amber is found. This pit amber has not been given the time for weathering to remove the crust. It has been dug from the earth by man instead. This crust or skin needs to be removed before it can be carved or utilized in the production of jewelry. Sea amber has been acted upon by the waves and the skin has been removed by nature.

Other pit ambers come from alluvial deposits. The amber was removed from the blue earth, transported to another location and buried under sediment. It was discovered by man in its new location and is mined there.

Inclusions

One of the most interesting things about amber is the inclusions found within it. Amber is the only gem that contains organic inclusions of fauna and flora. Amber fossils and inclusions are very unusual because they are three-dimensional representatives of life that once lived on Earth. Animals are often found in active poses that show prey and predator locked in an unending dance of death. We are the victor because we have a glimpse of the past that exists nowhere else. These relics are encased in their amber tomb and we have a window into our past.

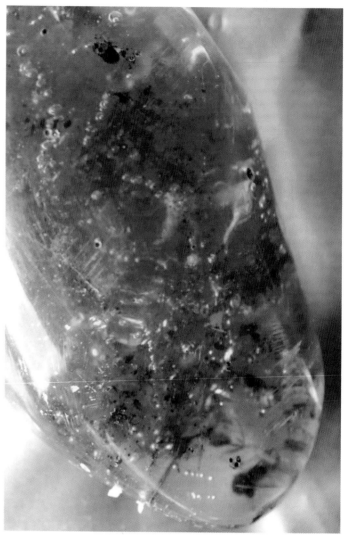

A termite colony was captured in this Dominican amber specimen. *Courtesy Patricio Jhonson, DR.*

Amber hearts with interesting inclusions are there to capture your heart. *Courtesy Amber Art, DR.*

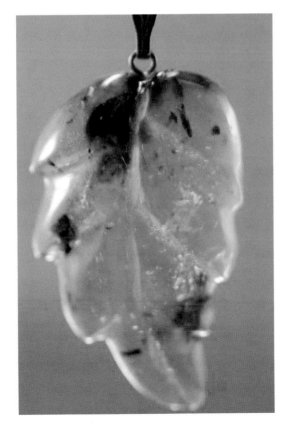

A beautifully carved leaf with plant inclusions resembles a fall leaf. *Courtesy The Sterling Rose*

Animal Life

The most common inclusions are flies, members of the Diptera family. These flies are also called fungus gnats. They lived in the fungus growing on the rotting vegetation found in amber forests. As small insects, they could have easily gotten stuck in the gummy residue and not been able to pull out. As the amber flowed from its source, layers of amber oozed over the captured prey and encased it. (www.ganoskin.com/orchid/archive/199702/msg00073.htm)

Amber gives us a time capsule of life in the ancient forests, but we must remember that it is an incomplete record. Many things were trapped in the sticky substance, but animals that were strong and healthy could have pulled out of the resin. It was only weakened animals and small insects and plant materials that were easily trapped.

Interesting inclusions create a blend of colors in this pendant. *Courtesy The Sterling Rose.*

Bubbles and plant matter provide a glimpse inside an amber specimen.

A re-creation of an ant entrapment in amber flows at the Amber Mundo Museum shows the viewer how it might have happened. *Courtesy Amber Mundo Museum.*

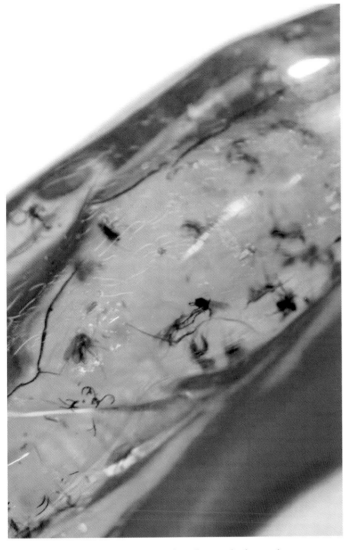

Flies are the most common insects found entombed in amber. *Courtesy Amber Mundo Museum.*

It is unusual but interesting to find the oldest living insect, cockroaches, in amber. Blattoidea, a species of ancient cockroach, has a fossilized amber record of every stage of pupation, but rarely is an adult found. The adults were large enough and strong enough to escape the amber trap. (www. ganoskin.com/orchid/archive/199702/msg00073.htm)

If you encounter amber with perfectly preserved animals and insects, be wary. Strong, fit animals would have escaped. Even those that were entombed must have struggled and the piece should show signs of this struggle preserved in the specimen. Broken legs, torn wings, and other signs should be apparent in real amber inclusions. These inclusions will never show vibrant color and a piece with any colorful inclusion should be especially questioned and will rarely be considered natural.

Occasionally some unusual things have been found in amber. What was believed to be a frog was discovered in Dominican amber. The bones were present, some were broken and most of the flesh had deteriorated. Upon close inspection, it was discovered that the frog had six or more legs. What paleontologists now suspect is that a bird feasted in the tree above the spot where amber was forming and the bird lost some of its prey. The catch dropped into the sticky resin where it remained. Several animals shared the trap. (www. ganoskin.com/orchid/archive/199702/msg00073.htm)

After amber resin was secreted, it would begin to go through a hardening process and a skin would begin to form on its surface. This partially hardened skin sometimes captured a glimpse of larger prey. The footprints from a cat and other animals have been preserved in Baltic and Dominican amber. The hair of mammals is frequently found preserved in amber. Feathers of birds have given us a glimpse at species that no longer exist. Even the spine and ribs of a mouse were found in Dominican amber. This species of mouse was not expected to have lived there 30+ million years ago and the find has made theorists rewrite their beliefs of the animal population that lived in the West Indian islands during that time period. (www.ganoskin.com/orchid/archive/199702/msg00073.htm)

Occasionally, unusual objects have been discovered. A small lizard was found in Dominican amber. These are very rare in European amber. Doctor Kosmowska-Ceranowicz has described a large set of mammalian molars encased in Polish amber. It is believed that the animal died with its face in an amber resin deposit and the teeth were completely 'amberised' while the terpines in the resin decayed the animal's jaw. (www. ganoskin.com/orchid/archive/199702/msg00073.htm)

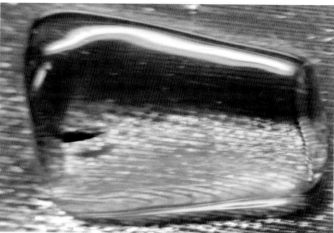

The typical cockroach found in amber is much smaller, like this specimen. *Courtesy Amber Art.*

Cockroaches are rarely found in adult form as they are normally strong enough to pull themselves out of the sticky substance. A specimen at The *Amber Mundo Museum* shows a large one captured in amber. The top picture shows the stone. The bottom one is a detail of the cockroach.

A detail of the piece helps us see the ancestor of man's nemesis, the roach. This is believed to be the largest specimen ever found containing the bug.

Plant Life

The study of the plant life found in amber is also fascinating. Seeds, spores and pieces of leaves have been preserved. Gymnosperm inclusions have come from fir, cypress, juniper, pine, spruce and arbor vitae. Angiosperm remains have come from fifteen different oaks, beech, maple, chestnut, magnolia, and cinnamon. Palms, ferns, mosses and other ground covers also occur as inclusions. Other inorganic inclusions such as sulfur and pyrite appear in amber specimens. (www.emporia.edu/earthsci/amber/life.htm)

Amber inclusions give us a perspective on ancient life and scientists' study of them is ongoing. DNA has been discovered in some insects and plants and there has been some conjecture that the DNA could be used to re-create life again. This scientific study inspired a famous movie, *Jurassic Park* ®. Science becomes fiction.

Baltic amber specimens with plant inclusions are of interest to both botanists and collector. Each of these specimens has an abundance of plant inclusions. *Courtesy The Sterling Rose.*

A closer look at one specimen shows a hollow where a leaf rested on the surface and another leaf trapped in the amber.

It has frequently been said that the number of plants and animals that are entombed in amber should be suspect. Baltic amber has about 1 inclusion in 1000 specimens and Dominican amber has about 1 in every 100 specimens. Buyers beware because not every inclusion is a natural one. There are companies that frequently sell fakes. Dealers are required by law to tell you if your specimen is real or not, but often neglect to do so. When you buy from a secondary source, they may not even know if the piece is real.

It was believed that inclusions only occurred in clear transparent amber; however, in March 2008, a group of researchers reported that they had discovered 356 animal species and bits of plants trapped inside 4.4 pounds of opaque amber. Their findings were reported in the journal *Microscopy and Microanalysis*. They used synchrotron X-rays to peer into the cloudy amber releasing high-energy light and produce images of the inclusions. Their findings are giving a new window into the past, as opaque amber makes up the bulk of amber retrieved today from the Baltic regions. (www. foxnews.com/story/0,2933,349028,0.html)

The two scorpions posed in a dance to the death are also Vietnam creations in pine resin. *Author's collection.*

Scorpions and bugs are a favorite of novice amber collectors. Scorpions are rarely discovered. Large scorpions, perfectly preserved like this one, would not have been trapped. The scorpion shown was created in Vietnam and is preserved in pine resin. *Author's collection.*

Colorful bugs found in amber are always fakes. The terpines and acids would have caused the color pigments to disappear over the centuries. Another Vietnam fake in pine resins. *Author's collection.*

Sometimes interesting specimens are created as artful displays. This creation was designed to resemble an ancient world with plants and scorpions. Baltic amber has been melted and re-cast with the inclusions. The material could test as amber, but the inclusions would be recent. An entomologist or botanist would recognize them as fakes, but a novice would find identification difficult. *Courtesy Joe and Eula Grove, the Coral Reef, New Port Richey, Florida.*

A detail look shows you the effort that went into creating the display.

Sometimes you come across a fake that almost no one would believe to be real. This bug in an ice cube should be immediately recognized as a fake, but I wouldn't want to find it in my drink. *Author's collection.*

Amber Sources

Ancient Origins

Amber has an ancient history that transcends time and brings a glimpse of the past into our present. The story is older than recorded history, beginning before man took his first steps on the Earth. Tens to hundreds of millions of years ago, amber's journey began.

However, it was with man's first steps that amber became an important component on the planet. No other stone has been as treasured and revered for its amazing beauty, unique properties, and mystery.

Three specimens of copal that closely resembles amber, but are much younger. *Author's collection.*

The differences in Baltic amber (seen in this picture) and Madagascar copal (seen next) are difficult to distinguish at first glance. *Author's collection.*

Amber was formed millions of years ago in forests that covered the Earth. The best known deposit was formed on a pre-Fennoscandian continent, a land mass that had broken from Pangaea, which contained all the land before it fractured and spread out across oceans to become the continents we recognize today. The northern parts of Europe, including Finland, the Baltic Sea, and northern Russia, were a continuous land mass during the Cambrian and Tertiary periods. From Spitsbergen on the north and Iceland to the west, it encompassed parts of Greenland and North America, the British Isles and northern France. (Rice, 1993, 16)

These amber forests covered much of the Northern Hemisphere at latitude 55 North and longititude 19 to 20 East, and extended from Europe into what are now Asia and North America. Even though the climate today is cold, the climate on the continent at that time is believed to have been warm and balmy. These temperate to sub-tropical forests supported a variety of life. Plants and animals grew there in profusion. The conifer trees that are believed to have produced amber existed with ferns, mosses, and other tropical fauna and flora.

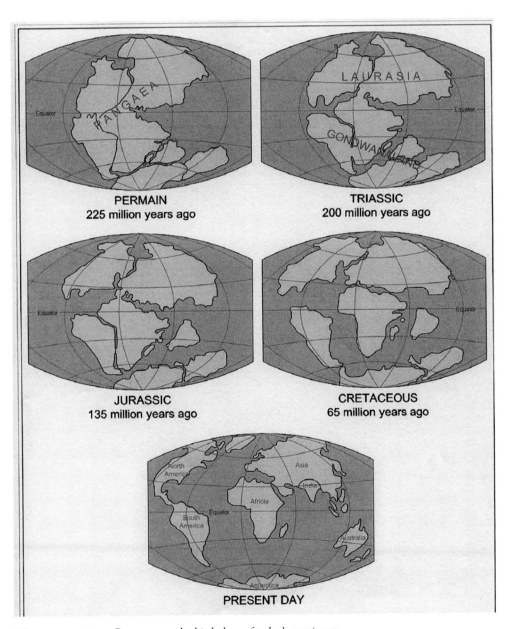

The super continent Pangaea was the birthplace of today's continents.

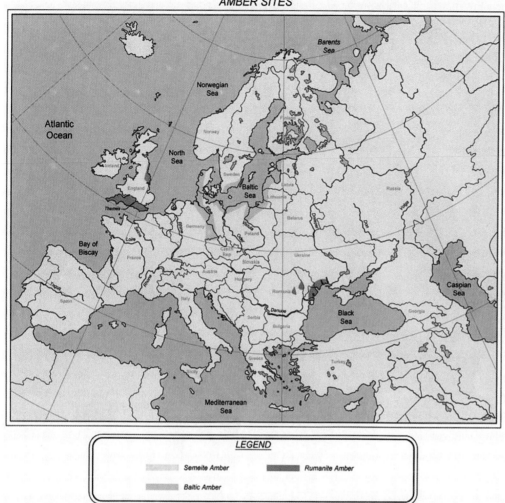

Baltic amber is collected from a region known as the Kaliningrad Oblast or Samland area.

LEGEND

Semeite Amber   Rumanite Amber

Baltic Amber

The artist's rendering of the amber forest depicts tropical vegetation growing alongside the amber producing conifers.

44

Looking at a cutaway rendering of an amber producing tree shows the locations where amber forms within the tree as well as flows from the tree.

When the now-extinct conifer trees, thought to be Pinus Succinfera in the Baltic regions and other species in the rest of the world, suffered from disease, were damaged by the elements, or were attacked by insects, they secreted a gummy resin to seal their wounds. The forest was under constant attack by the wind, rain, and lightning and the trees produced enormous quantities of resin. Some was preserved within the structure of the trees; other resin was stored under the bark, and even more was released from the trees and fell to the ground.

At the Amber Mundo Museum in Santo Domingo, Dominican Republic you can see a display depicting how amber oozed from a damaged tree. Consider how easy it would be for anything to stick to this gooey secretion. *Courtesy Amber Mundo Museum.*

As the trees grew old and succumbed to disease, they fell to earth where they were buried under layers of sediment. Gigantic upheavals in the earth caused areas of these forests to sink and be submerged in water, causing destruction of the forests. (Sinkankas, 1996, 596) The trees decayed, leaving just the clumps of resin. This tree resin (a mixture of various organic compounds, including succinic acid and succinic resins) was buried under layers of debris. A *blue earth* containing clay over thirty feet thick in some regions, was deposited over the amber. It was heated and compacted by enormous pressure and metamorphosized over the centuries. The resin lost most of its more volatile compounds, polymerized, hardened, and created this amazing amber "stone." The change from resin to mineral was complete. (Simon & Schuster, 1986, p.308)

As rivers were formed and changed course, the remains of these forests were transported toward the seas. The continents continued to drift and break apart. Glaciers covering this region over two million years ago caused the earth's crust to move and some of this buried treasure was brought to the surface. Some of the amber was washed out of the blue earth and was transported by rivers and tidal forces to coastal locations. However, some amber remained in pit deposits deep within the earth. When it was discovered, man wanted to release it from the ground and enjoy its unique attributes. The best known ancient amber sources, and the ones with the longest history, are in the present-day Baltic countries of Latvia, Lithuania, Russia, Poland and Denmark.

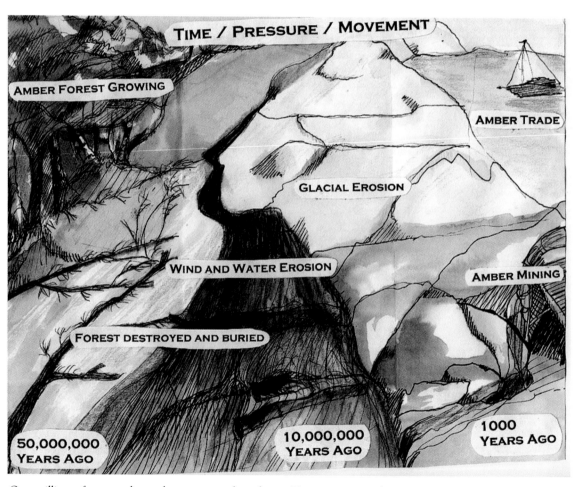

Over millions of years, amber undergoes tremendous change. Time, pressure, and erosive forces change the tree resin into copal and eventually into amber.

Most amber has been found on shorelines in northern climates near areas where the trees once grew. Some is dug from pits found under the blue earth. In the northern Baltic regions, where amber was first found, it was loved and treasured as the *gold of the north*. This *gold* washed up on shores in the Baltic regions after heavy winter storms. Baltic amber was among the earliest gem materials utilized by man. It had several unusual properties that made amber in the Baltic especially valuable to the cultures and it was loved and even worshipped above most other stones. Pieces of amber, from slivers to larger stones, were sought.

The largest known amber deposit in the world lies beneath 100 feet of sand near Kaliningrad, Russia, and has been worked for over one hundred years. (Schumann, 1997, p. 228) Other amber deposits are found in delta regions across the Earth where they were transported by water and wind.

Most amber that is newly found today is in small clumps weighing up to twenty-two pounds. They have been turned into a variety of objects in both the present as well as the past. Sea amber of all sizes is especially solid, easy to carve, and perfect for turning into jewelry and religious objects.

The small amber nuggets were retrieved from the amber mines in the La Cumbre region, D.R. *Courtesy The Amber Factory, La Cumbre, DR.*

Historical / Amber Timeline

Million yrs.	ERA	PERIOD		EPOCH	EVENTS
0.01				Holocene	**Neolithic: New Stone Age to PRESENT**
1.6			Quaternary		African Copal & Kauri Gum formed
					Paleolithic Old Stone Age Amber used by man
5				Pleistocene	Humans Established / Ice Age
				Pliocene	Amber transported & re-deposited by nature
			Neogene		
24	Cenozoic			Miocene	Blue Earth formed over succinite amber
		Tertiary			Dominican & Simetite Amber Formed
37				(H)ologocene	Drastic Climate Changes
			Paleogene		Extinction of Amber Forrest
				Eocene	Burmite & Rumanite Amber formed
					Baltic Amber Forrests Decline
				Paleocene	
66					
114		Cretaceous			Super Continent Pangea Breaks up
					Succinite Amber formation begins
	Mesozoic	Jurassic			mammals
208					birds
		Triassic			Pangea being torn apart by continental drift
245					dinosaurs
		Permian			
286					insects
320		Carboniferous	Pennsylvanian		reptiles
360			Mississippian		Pangea, the super continent existed
	Paleozoic	Devonian			amphibians
408					
		Silurian			land plants
438					
		Ordovician			
505					
		Cambrian			
540					vertebrates
2500					
		Precambrian			
3800					beginning of life

The journey of life on earth has taken many turns. Amber appeared before humans, and has been their constant companion ever since.

47

These beautiful cleaned and polished amber gems are available for sale in Santo Domingo, D.R. The inclusions add to their beauty and value. *Courtesy Amber Art.*

A large yellow amber chunk weighing about 6 pounds., found in its original rock matrix, is on display at Amber Mundo Museum, D.R. *Courtesy Amber Mundo Museum.*

Modern Origins

Amber is a universal gem material and has been used and loved by most cultures around the world. Either it was discovered by indigenous people or it was traded to them as items of beauty, magic, and curiosity. This gem continues to be a valuable commodity today.

When people think of amber, they most often think of the beautiful gem from the Baltic region. It has the richest history and the best reputation. However, there are other locations where amber can be found. These finds are not as well known, but they are equally fascinating.

Fushun Amber

An unusual amber is found in China. Fushun amber is found in the coal mines near the city of Fushun. When the coal is extracted, the amber is separated from the coal and carved into small statues. This amber is a mixture of yellow and black. Coal is not a normal habitat for amber, and gives this gem an unusual history.

Burmite Amber

Burmite, amber found in Burma, was exported from what is now Myanmar to China from the first century A.D. and was utilized in jewelry and carvings. The cherry-red color made it an especially good match for jewelry containing ivory and jade. This amber was also a desirable material in Britain during the Victorian age, when Baltic amber became scarce. The cherry-red beads became popular, starting a fashion craze among fashionable women.

Burmite was mined from 1898 until 1941 and, according to records of the Geological Survey of India, 82,000 kg of this material was excavated. However, only one collection of Burmite, used for scientific studies, was known to exist at that time. It consisted 117 pieces which contained 1200 organism inclusions. This collection is located at the Natural History Museum of London. Mr. R.C.J. Swinhoe of Mandalay assembled this collection between 1915 and 1916 and proceeded to send it to a renowned entomologist in the United States, Mr. T.D.A Cockerel, who published thirteen papers on the collection and described forty-one new arthropods. He further used his understanding of flora and fauna to determine the amber to be from the Cretaceous age, even though the surrounding material was of Miocene origin. The Cretaceous age is one of the most significant periods in the study of terrestrial life and was not well understood until that time. These specimens provided a time capsule for studying early life on our planet.

Mr. H.L. Chhibber visited the area where the amber was excavated in 1930 and reported that the amber occurred in thin lignite seams among clays and shale. He further noted that the larger and more transparent pieces were found at a depth of 10 to 15 meters.

Amber of Burmese origin is unique. It is the only fossiliferous amber deposit from this time period found in southeastern Asia. Other deposits of the Cretaceous period are found in the Northern Hemisphere. How it arrived there is another mystery in the world of this amazing stone.

The excavation of amber in this area was interrupted by World War II. In 1999, a Canadian gold mining company resumed mining this beautiful gem material. (Grimaldi, 1-5) Today, this amber is difficult to attain because of the political climate in that region.

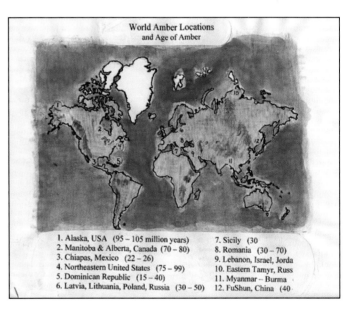

World Amber Locations and Age of Amber

1. Alaska, USA (95 – 105 million years)
2. Manitoba & Alberta, Canada (70 – 80)
3. Chiapas, Mexico (22 – 26)
4. Northeastern United States (75 – 99)
5. Dominican Republic (15 – 40)
6. Latvia, Lithuania, Poland, Russia (30 – 50)
7. Sicily (30
8. Romania (30 – 70)
9. Lebanon, Israel, Jorda
10. Eastern Tamyr, Russ
11. Myanmar – Burma
12. FuShun, China (40

The best known amber is found in the Baltic regions. The second largest amber supply is from the Dominican Republic, but there are amber finds all over the globe. Most amber is found in the middle latitudes. This map shows the locations of these and other amber finds and gives the age of the amber. Copal is found outside of these latitudes as climate has changed since the initial formation of amber millions of years ago.

Dominican Amber

The second largest supply of amber discovered today comes from the Dominican Republic. Dominican amber is believed to be a resinous product produced by the Hymenea, a leguminous tree once found there. It is mined in three regions on the island: the La Cumbre region where amber 30 to 40 million years old can be found; the area around El Valle where slightly younger amber is mined; and the Baya-guana areas where some amber and some copal are found. Most of the copal is 15 to 17 million years old. Amber from the Dominican Republic has become most sought after by entomologists and paleobiologists because of the types and number of inclusions found within it. As most Dominican amber is naturally clear, the specimens within it are easy to observe and identify.

The amber miners in the Dominican Republic live simple lives. Their homes are clustered near where they work and they are able to grow most things they need. Coffee plants, bananas, oranges and even cacao (the cocoa bean) are growing near their homes.

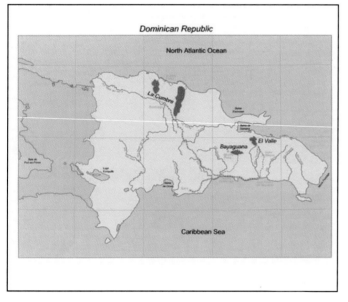

The Dominican Republic has three primary regions where amber is found: La Cumbre, Bayaguana, and El Valle. The mountainous areas around Santiago and Puerto Plata form the La Cumbre area. The Bayaguana mines are near Monte Plate. The El Valle mines are in the remote Hato Major region.

The Dominican mines are located in remote mountainous areas in a setting resembling those shown in the movie *Jurassic Park®*, which made Dominican amber famous. Roads into the region are susceptible to landslides and ongoing construction attempts to keep the roads open. The land where the amber lies is owned by families that have controlled it for centuries. Some of them are mined by 'coffee pickers' who work for the patron during their off seasons to make more money. Others spend their entire lives working the mines. They live in small huts with tin or palm leaved roofs. Few would change their chosen occupation as they were brought up in the amber mining business and know of no other job that would offer them pay of about $10.00 / day, which is about twice the average salary of workers in the sugar cane and coffee fields.

When you think of *Jurassic Park®*, the movie, you can imagine the beauty of the land where Dominican amber is dug. Misty mountains and lush vegetation are everywhere.

Near the top of a hillside holding the Amber Castino mines, you see the tools that are regularly used. A wheel barrow lies in wait to move the dirt excavated from the mine. Small trees have been stripped and lay in wait to be used to brace the ground around the mine.

A shovel, pick, hammer and spike lie in the dirt on a step leading down into the mine. These are the miner's primary equipment.

This mine descends about twenty feet straight down and then small tunnels recede into the mountain where the amber is mined. The tools and a candle are the miners' only companions as they work for hours at a time inside the mountain.

In the La Cumbre area, holes are dug into the clay and limestone cliff and pits are dug into the ground. The amber is removed by hand with a hammer and chisel. The mine opening may be as wide as 10 feet across, but at the bottom of the shaft, 20 to 25 feet below ground, smaller shafts take off in a variety of directions. The fox holes at the bottom are barely wide enough for a man to crawl. There is no light and the miners utilize candles to light their way. The miners spend their days underground where they lie in the earth or crouch as they chip amber out of the rock.

In the El Valle area amber is closer to the surface and miners create caves in the hillsides with a machete to remove the material from its matrix.

The Bayaguana area supports a combination of methods. Amber in the Dominican Republic belongs to the landowner, and the miner is paid for retrieving the stone. If no amber is discovered, no payment is made. If amber is discovered, a find could pay for the simple needs of the family for a month.

Patricio Jhonson shows off a large piece of amber extracted from one of her mines in the El Valle area.

This beautiful grotto is really one of the Jhonson's mines.

Mr. Jhonson uses a machete to cut into the clay and retrieve amber.

This hollow is another mine area on the Jhonson's property.

Along the roads in the mining areas, small shops are located that sell the amber mined there. The shop owner is probably the landowner, and most enjoy taking you on a tour of their mine. These same people can also be the artisans who create the jewelry and artworks found in their shops. Do not expect to get a fantastic buy here, as they are well aware of the value of amber in the shops in the larger cities and do not discount much.

Amber from the Dominican Republic is found in a wide variety of colors. Most Dominican amber is yellow, gold, orange, or brown. However, there are also ambers in red, blue, black, and green hues. The blue amber from here is especially beautiful and expensive.

Dominican amber is considered to be clearer than most amber, and it has more inclusions. Because it is naturally clear, it is not given special heat or chemical treatments like the Baltic amber we find today.

The Amber Factory is a larger shop selling amber and Larimar products from the Dominican Republic. Larimar, a stunning blue stone, is only found in the Domnican Republic.

Amber Castillo, a small shop and mine, is located alongside the road. The people are very friendly and helpful.

Mexican Amber

Amber is also found in Chiapas, Mexico. Chiapas is the southernmost state in Mexico and lies at the entrance to the Yucatan. Mayan people lived in this area as early as 36 BC and used the amber in religious rites. Amber from this region is thought to be between 22 and 46 million years old. It too is believed to be a resin produced by the Hymenaea Leguminoseae tree. The rare red amber can be found in this region; however, most of the red amber coming from the mines crumbles into a powder and only a few pieces are good enough for jewelry.

Other Major Amber Deposits

Amber is usually found in select coastal areas around the world. The Baltic area is the best known but there are many other areas where it can be found. Greenland is a prolific area for amber, as is Puerto Rico.

Manitoba, Canada, produces amber called *cedarite*. It is found in an area that was once considered to be a delta. British Columbian amber is associated with coal deposits found along the Nechako River.

In the continental United States, amber is found in Massachusetts, New York, New Jersey, Maryland, North Carolina, Tennessee, Mississippi, Arkansas, Kansas, Texas, South Dakota, Colorado, Wyoming, New Mexico, and California. The largest amber deposits in the continental United States are in openpit clay mines in Arkansas. The best known American amber comes from abandoned pit mines in New Jersey.

Amber is found also near Point Barrow, Alaska.

Chiapas amber is found in Chiapas, Mexico. Chiapas is located at the entrance to the Yucatan near the Mayan ruins of Palenque. The Mayans used amber for incense and treasured the stone as well. This large piece of red amber is rare as most pieces of red are very small. *Author's collection.*

A polished specimen of Chiapas amber shows both red and amber coloration. *Courtesy Charles Albert jewelry.*

Yellow amber is found in larger pieces in Chiapas and this specimen makes a great pendant. *Author's collection.*

Amber in History

A History of Amber

Amber has been prized since the Stone Age and was given special names by cultures that encountered it. Persians called it *kahraba*, "the attractor of straw," because amber exerts an electrical force when rubbed briskly with material and attracts wood chips or dried grasses. Greeks called it *electron*, "sun-gold," and knew of its electrical powers. Germans referred to it as *bernstein*, "burning stone," because it would burn, as would few others. The name we know descended from the Arabic *al-anbar*. What many people consider the most beautiful name was a Tibetan word *p/-she*, or "perfumed crystal." The beauty of the stone, its refreshing odor, and its unusual characteristics made it desirable.

A large Baltic amber lump shows the age cracks amber gets from being subjected to weathering over a long period of time. *Author's collection*.

This beautiful Baltic amber specimen shows what is left when debris deposited in the amber at its creation decayed and left holes for us to discover. *Author's collection*.

Once it was carved into animal shapes or had designs incised onto the surface, amber was attributed with even more power. The natural depressions found in rough amber were left intact in early pieces because they were believed to contain the spirits of animals. (Elsbeth, 2002, p. 138-139)

Amber turtles were representations of animal life that was found in the area at the time of the carving. *Courtesy Amber Castino.*

Amber dragons were carried by men to stimulate their virility. *Courtesy Amber Mundo Museum.*

Carvings of fish, frogs and rabbits were carried or worn by women desiring to conceive children. *Courtesy the Amber Factory, La Cumbre.*

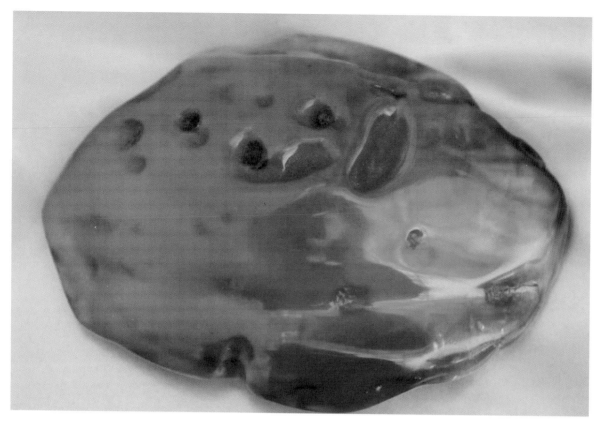

Amber with natural depressions was believed by ancient man to hold animal's spirits. This cloudy amber specimen must hold the spirits of animals not desiring to be seen. *Courtesy private collector, Renningers' Antique Show.*

A family of spirits could be housed in this Baltic specimen. *Author's collection.*

Can you imagine a caveman picking up a stone, rubbing it on his clothing to clean it, and discovering that it attracted small seeds, feathers, and even dried grass when it was placed down again? They had no knowwledge of static electricity and would have been amazed and probably frightened by its power. Can you imagine this same caveman fearfully throwing the stone into the fire to destroy it and discovering that it burned or melted and gave off a very pleasant piney scent? What stone burns? Very few. It would be as amazing for them to discover that this same stone, when thrown into the Baltic Sea, would not sink into the salt water as expected, but float. And some of these stones were filled with plants and animals that gave the Stone Age people a frozen picture of life they might have glimpsed around them. What a treasure. Reverence for the stone grew with each new discovery.

A cave man picking up this amber would be amazed at what he saw. Would the spider protect him from future bites or would it be a warning of impending danger?

This beautiful specimen of Chiapas amber would make an interesting gazing stone for prehistoric people.
Courtesy Charles Albert Jewelry, FL.

Archeologists have determined that the people of the Upper Paleolithic Period (25,000 – 12,000 B.C.) decorated themselves with shells, bone, teeth, and unique stones. It is believed that as early as 13,000 B.C. ancient man utilized amber to create amulets, pendants, and beads and that amber was one of the most sacred stones of this period. It was carried by hunters, drilled with primitive flint knives, and strung for wearing to ensure safe and prosperous expeditions.

Notice the age cracks and coloration on this lovely multi-tone amber necklace. *Courtesy The Sterling Rose jewelry.*

Joe Grove displays some of the amber beads he has for sale at a G&LW wholesale jewelry show. *Courtesy The Coral Reef.*

These beads are ready to string. From left to right: blue Dominican amber beads and pendant, African copal beads, and synthetic amber beads. *Courtesy The Sterling Rose jewelry.*

These pink-tone plastic beads may resemble amber, but are worth a fraction of amber's value. *Courtesy The Sterling Rose jewelry.*

Initially, amber was collected in its natural state. Unworked amber was found in caves in the high Pyrenees Mountains of France and other locations in southern Europe where it had been placed by ancient man as early as 8200 B.C., making it perhaps the oldest known collection of amber gathered by mankind. However, David Grimaldi stated in *Amber, Window to the Past* that worked pieces of amber in the form of beads were found in southern England in Gough's cave, dating from 11,000–9,000 B.C. Other researchers place the oldest worked amber at about 7,000 B.C. in Denmark. A pendant depicting four human figures with striped patterning was discovered there in an ancient bog, an environment that kept the amber in a remarkable state of preservation. (www.gplatt.demon.co.uk/abrief.htm)

Amber was collected by many cultures and some was carved into beads, talismans, and other ornamentation found in archeological deposits around the world. The Aisti people of the eastern and western Baltic region, the ancestors of the Latvians and Lithuanians, were some of the first amber gathers. They use amber for personal decoration and traded it with other nearby cultures. Amber's ability to be easily worked with ancient tools and its mystical value made the amber trade spread.

Artist rendering of Stone Age and Bronze Age amber figures and amulets found in the Baltic region.

This copy of an ancient amber artifact shows us how the carvers worked their magic carving with Stone Age tools. Amber ages over time and looses its luster. If the piece were polished, it would be brought back to its original luster, but would loose its antique value. *Courtesy Amber Mundo Museum.*

Discs of amber, found in several archaeological excavations of about 8000–4000 B.C., were carved with designs believed to have religious meaning. The centers of each disc were removed, the surfaces were polished, and double rays of dots were carved to form a cross. It is believed that these indicated sun worship and the discs were symbols of a sun-wheel cult. (Rice, 1993, 28)

In early Europe and the Mediterranean region, amber was a principal bartering tool. Amber ornaments were discovered off the continent in mound tombs of Great Britain, especially around the Stonehenge area. The amber trade extended as far as central Russia, Finland, and western Norway as early as 3000 B.C. Amber ornaments were found in Egyptian tombs dating to 3200 B.C. (Rice, 1993, pp.32-33) Amber beads dating from 2000 B.C. have been found in Crete and England, while amber pendants and beads dating from 3700 B.C. were discovered in Estonia. When these treasures have been tested to discover where they originated, most were of Baltic origin.

Phoenicians were probably the first sailors to trade in amber around the Mediterranean Sea and along the Atlantic coastline of northern Europe. Spectroscopic analysis of amber beads found in graves of Mycenaean Greece and the fabled city of Troy supports their involvement in early amber trade. (Rice, 1993, 38) They carried it for sale to Northern Africa, Turkey, Cypress, and Greece. It is believed that they obtained amber as a trade good and exchanged it for bronze or tin between the thirteenth and sixth centuries, B.C. It is further believed that they were unaware of its true origin; but once they discovered it, they went to great lengths to hide amber's origin from their buyers. (Rice, 1993, 38) In her book, *Jewels and Gems*, Lucille Saunder MacDonald retells a story that Phoenician traders told others to protect their secret about amber's origin:

Now that the Phoenicians had seen the amber gathered from the sea, they determined to keep the secret for themselves and thus guard the lucrative trade. When the fleets returned to Syria, many were the tales told of perils to the north, of lodestones which would draw the ships to destruction on hidden reefs, of whirlpools which would suck them down to the bottom of the ocean, of witches who enchanted men by turning them into beasts, of terrible sea serpents, and awesome monsters. So well did these ancient sailors spin their yarns that for many centuries afterwards mariners feared these mythical perils." (gplatt.demon.co.uk/abrief.htm)

These stories recall other tales told about the same time. Homer's *Odyssey* recounts similar perils. Ancient sailors were afraid to sail far from home because they feared falling off the edge of a flat earth or being eaten by sea monsters. Such tales were effective in protecting ancient trade routes.

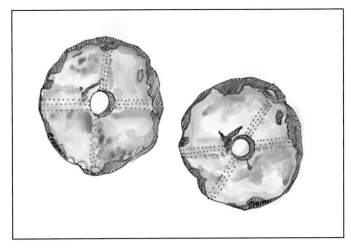

Artist rendering of amber discs from Sambian Promontory circa 2500 BC. These discs were believed to represent the sun god and were carried by the Sun God cult.

Another remarkable amber find, the Hove Cup, was recovered from a burial mound in Hove, Britain. This cup measures 6.4 cm. tall and has an opening of 8.9 cm. across. It is one of the most famous and best preserved pieces of amber dating from around 1,500 B.C.

The Celts dominated Europe between 400 B.C. and 30 A.D. and their culture permeated Europe, Italy, and the Balkans. Trading in and using amber was important. Artisans from Celtic regions created some of the most beautiful amber products and transported them afar. As they traded, they shared in a cultural exchange that helped to spread civilization.

Artist rendering of the 3500 year old Hove Amber Cup recovered in a burial mound in Hove, Britain.

Gemstones set in rings were a late addition to human adornment, but carved stone rings were worn earlier. Rings made entirely of gold, silver, iron, carved amber, and ivory were among the early Egyptian, Greek, Roman, Etruscan, and Middle-Eastern cultures. They were plain bands in the beginning, and gradually became ornate, eventually becoming signet rings. Being easy to carve, amber was a natural material for this progression.

Artist rendering of a first century Roman amber ring. Early rings were created as simple bands, but later artisans produced beautifully carved pieces.

Beads and necklaces were equally important. Homer's *Odyssey* has one of the earliest mentions of amber in literature. It states that Eurymachus was given an amber bead necklace. (Conway, 1999, 36) The gift reflected the high value ascribed to this treasure. Amber has continued to be a desired component of jewelry throughout the ages.

Roman jewelry utilized Baltic amber, and beads made from amber and jet were highly prized. Pliny, the Roman historian, wrote that an amber figurine of any size was worth more than a slave, even a male slave in his prime. During the reign of Emperor Nero, a knight was sent to the far north to locate amber and bring it back to Rome to decorate the gladiator's arena. This knight brought back such an immense quantity of amber that the nets used to protect the observers from the games were studded with it, animals used in the

This cognac amber necklace with round beads and a lump pendant was for sale in the Dominican Republic. *Courtesy Turi Gift Shop.*

games had buttons of amber adorning them, gladiators were decorated with amber, and even the sets used to stage the games were made with amber. (www.gplatt.demon.co.uk/abrief.htm) Roman women with hair the color of amber were considered especially beautiful. By the first century A.D., Rome had fully established amber routes from the Baltic coast to the Black Sea and Mediterranean Sea, where it was shipped to Greece and Rome. From Rome, it was transported by land and water to Amsterdam, Antwerp, Bruges, Köln, Lubeck, Nuremburg, and Venice.

Faceted Russian amber beads are strung with 14k gold beads to give a glow to this prized necklace. *Author's collection.*

Natural amber discs of assorted shapes and sizes create a beautiful necklace. *Courtesy Turi Gift Shop.*

This necklace is a wonderful reminder of a trip to the Soviet Union and Russia. *Courtesy Donna and Paul Silagi.*

During the Bronze Age, amber was made into beads and jewelry. The Tara Brooch, one of the most famous Celtic brooches of the sixth century, was crafted from silver and set with amber and colored glass studs. Women of the seventh century wore lengths of amber beads across their chests. Religious jewelry made from amber was in demand. In Petrus Christus's painting of St. Eloy, the patron saint of goldsmiths, completed jewelry, including brooches, amuletic pendants, and a box of rings, is displayed on rolls of parchment. Next to these are packets of loose gemstones, pearls, pieces of rock crystal, and large amber beads. (Phillips, 1996, p. 58) This confirms the belief that amber was treasured as the gold of the north.

In the earliest times, amber was the property of the finder. As the amber trade became a lucrative business, dukes, kings, and Teutonic knights attempted to control its collection and sale. Fishing rights for amber were awarded and revoked by Amber Lords as early as 1264 A.D. Amber riders scoured the shoreline on horseback looking for amber that had washed up. They used poles and nets, called amber catchers, to retrieve the treasure.

In 1308, Teutonic monks gained control of the amber trade. They wanted to control Gdansk, which was a central location for harvesting and mining amber. As their powers grew, the monks became bold. They entered Gdansk, murdered most of the townspeople and burned their homes. The Gdansk Massacre, as it was called, caused the amber trade to collapse. (www.khulsey.com/jewelry/kh_jewelry_amber_mining.html)

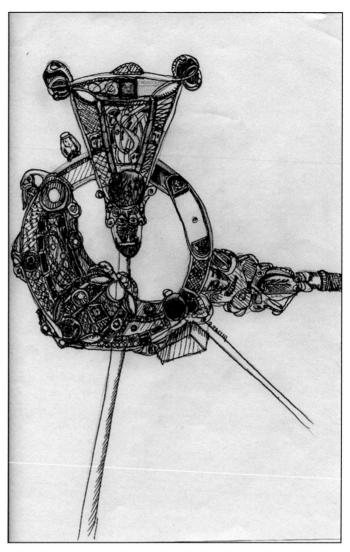

The Celts were master silversmiths and talented in working with amber. This artist rendering of the Tara Broach from the sixth century portrays some of this excellent workmanship.

Artist rendering of paternosters, prayer beads, showing amber beads strung in a loupe with a tassel at the end. Prayers were offered as the beads were pulled along the cord.

Amber was collected by Baltic Sea divers the way sponges were collected by Greek divers. Matt, with the St. Nicholas Boat Line, shows today's tourists in Tarpon Springs how sponges are pulled from the Gulf of Mexico.

In 1725, the government hired two professional divers from Halle to find amber that was too deep to be reached with forks and probes. This venture was a financial failure. In 1869, diving was again attempted and yielded a rich amber harvest. (Rice, 1993, pg. 69) These divers first dressed in woolen garments to keep themselves warm, then donned an "Indian-rubber suit" with brass fittings. Over this was placed a brass and glass helmet weighing close to thirty pounds with three glass openings that allowed the diver to see in three directions. The helmet screwed onto the brass fittings and was attached to an air pump in the boat. A rope was securely attached around the diver's waist to retrieve him in case of an accident. Shoes made of wood and lead were worn to assist the diver to sink to the bottom. Other weights were placed over his shoulders allowing him to stay on the bottom for up to five hours. Usually, seven men served as the crew: two were divers, two pairs worked the air pumps, and an overseer watched to insure that the catch made it home.

Before descending into the waters, Matt checks his suit to insure there are no leaks. His hose and rope are securely attached and functioning.

Matt is fully suited up and ready to go overboard in search of his treasure.

During this period, amber was generally collected under the supervision of a Beach Master or Beach Rider. When amber fishing was not properly supervised, thieves took advantage; but when discovered, they and the persons permitting the illegal collecting were hung. Peasants possessing pieces of amber that were not considered religious articles could also be killed just for possessing it. Coastal villages near where amber floated on shore were likely to have permanent gallows. These served as warnings to the public and were waiting to deliver justice to amber thieves.

Amber that washed onto the shore was collected. Tools were designed to retrieve the amber found in the seaweed and under the waves. Wooden spades, hooks, and poles were used to dislodge amber from the sand under the Baltic Sea. Always the masters were there protecting the treasure. As the rules and supervision became stricter, the amber trade declined.

Artist rendering of amber fishermen searching for amber near the shoreline. The authorities stood by to insure that all the catch was recorded. Gibbets, gallows, were always ready to provide justice to anyone found guilty of stealing. Rendering recreated from an old copper engraving by Wagner, first drawn in 1774.

The amber trade made a slight comeback during the late thirteenth century when paternosters, sets of beads strung to facilitate prayer, became very fashionable. Portraits from this time often show them hanging from a girdle or draped around a wrist. Some were simply made of beads of unremarkable origin strung in groups of ten, called decades. Some were strung in a continuous loop while others were left open with fancy tassels at the ends. However, there was demand for the best and these prayer beads were made with Sicily's coral, Baltic amber, Whitby jet, German agates, Venetian glass, and even gold. (Phillips, 1996, 73) Artisans risked their lives, if they were not authorized, to create these ornaments.

Artist rendering of a monk "bidding his beads", offering prayers to the Virgin Mary, as others look on in wonder.

Amber made a more significant comeback when the artisans united into guilds. The first amber guild was formed at Gdansk, Poland, in 1477. Casmir the Jagiellonian gave the city land containing amber deposits and amber production returned. Guild artisans created rosaries and works of art from raw material supplied by the Amber Lords who sold or gifted them to the aristocracy and the clergy.

Amber made contact with the New World as exploration began. The diaries of Christopher Columbus make reference to the first documented amber in the New World. Columbus occupied Hispaniola in the 1490s and wrote that he presented a Taino chief, a local indigenous leader, with a strand of European amber and received shoes decorated with Dominican amber in return. Dominican amber mines were never exploited, since the Spanish were more interested in the gold discovered there. (amberlady.com/article.htm) Dominican amber was forgotten until a recent revival of the mines. In the New World, amber was also utilized by Aztec and Mayan people who carved it and used it as incense. (www.madehow.com/Volume-7/Amber.html)

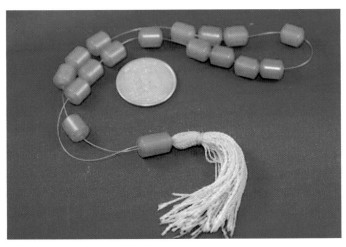
These middle eastern prayer beads were restrung in a loupe on mono-filament line with lots on space to move the beads. *Author's collection.*

Control of the amber trade continued, and in the 17th century, fishermen were required to swear to the Amber Oath. They had to denounce amber smugglers. Searching for amber, which was forbidden for most. But some amber gatherers were required to work in the harsh conditions of northern winters to harvest it. In a drawing from the first known book about amber (Hartmann, 1677), two amber fishermen are pictured in leather armor gathering amber with nets during the winter. A fire depicted in the background was utilized to melt the ice off their clothing as they toiled to earn a living. Harvesters often gave their lives for amber.

Amber's high value demanded payment in equally valuable products. Some amber fishermen were paid in salt, and the pay rate was weight for weight. (www.emporia.edu/earthsci/amber/recover.htm) Salt, as a requirement for life, had high value at the time. Since there was no refrigeration, salt was used to preserve foods as well as improve its taste. Other amber fishermen were allowed to keep two thirds of their amber catch as payment; the other third was turned over to the controlling body.

Amber's popularity steadily increased during the nineteenth century and the amber road became firmly established.

Artist rendering of amber fishermen wearing leather armor at dawn in early spring, 1670s. The Baltic Sea's cold sting could only be handled by escaping to a warm fire on the shore. Re-created from an illustration in the first book of amber by Hartman.

A Baltic amber clump shows the flow lines characteristic of natural amber. *Author's collection.*

Amber became an especially desirable material for memorial jewelry after the death of Britain's Prince Albert, in 1861. Amber met the mood of the people, their dress, and their desire for lightweight jewelry. Amber was not considered flashy and therefore was proper adornment for mourning. When demand outgrew the supply, amber became scarce and substitutes for it had to be found. Copal, a similar resin but much younger in geological age, became one of the substitutes. Plastics were also created around 1900 to simulate amber. (Parkinson, 1988, p. 227)

Amber beads and pearls continued to be fashionable during the Art Nouveau period, as artist Mogens Ballin (1871–1914) established a silver workshop in 1900 in Copenhagen. There he produced jewelry items with powerful organic fluid forms. One of his workers was Georg Jensen, who established his own silversmithy in 1904 and developed jewelry-making skills to an even more refined level. He created bezel-set cabochon jewelry in silver, copper, brass, and pewter that was set with semi-precious stones, such as lapis lazuli, amethyst, amber, and agate. (www. waddingtons.ca/pages/home/index.php?c=feat_ref/feat_08_99.php)

Columbian copal resembles amber and even has beautiful inclusions, but is much younger in age. *Courtesy Charles Albert jewelry.*

In the 20th century, amber waxed and waned in popularity, as it has for centuries. Amber jewelry became less fashionable as plastic jewelry became more popular. Celluloid, Bakelite, Lucite, and other plastics were in high demand during the early to mid-nineteenth century, causing amber to take a back seat. As plastics became less favored, amber has had a resurgence of popularity.

Plastics and other substitutes were used during the Victorian Age as substitutes for amber. This early celluloid and Lucite pin was designed to resemble amber and ivory. *Author's collection.*

Bakelite became a popular substitute for amber and eventually became desired in its own right. These Bakelite beads resemble the colors of amber, but clearly aren't amber. *Author's collection.*

Timeline for Amber and Humanity

1,000,000	8000	7000	6000	5000	4000	3000	2000	1000	BC/AD	1000	2000

Paleolithic Period Man Originates

Amber beads deposited in Goughs' Cave, Britain

Amber chunks deposited in caves at Hautes Pyrennes

Amber Trade Begins

Amber in Egypt, Troy & Mycenae

Hove Amber cup Britain

Amber recorded in China & India

753 BC Rome

1283 AD Teutonic Knights

1400 AD Amber Guilds form in Europe Columbus receives Taino amber from D.R.

500 AD Dark Ages

Victorian Age Amber is fashionable

1970 Dominican Government Promotes Amber

Amberif

The general timeline portrays ambers close link to humankind and some of the major amber finds. Many artifacts have been recovered from the Stone Age to the present.

The Amber Road

As desire for amber circled the world, a need to transport it as a saleable trade good developed. As the markets grew, so did the transportation system. An Amber Road was established.

This transcontinental trade route coincided with the Baltic Bronze Age, when amber was traded for tin and copper. Amber was initially traded to central European cultures, then sent to Mycenaea and Greece via the Amber Road. According to Marija Gimbutas, a Lithuanian anthropologist, it began around 1600 B.C., according to finds in Mycenaean tombs. (Rice, 1993, 36)

The Amber Road was a system of waterways and ancient highways that for centuries spanned Europe, expanded into Asia, and returned. It also stretched from northern Europe to the Mediterranean Sea. Amber was transported from the North Sea and Baltic Sea by numerous rivers and over land to the rivers of Italy and Greece, on to the Black Sea, and across the Mediterranean Sea to Egypt.

In Roman times, one route moved amber from the Baltic coast into Prussia, through Bohemia, and on to the Adriatic Sea. Here it was sent on to the temple of Apollo at Delphi as an offering to the gods. Earlier, amber had made its way into Egypt where Tutankhamen had it placed among his burial goods. Amber that made its way to the Black Sea would continue on to Asia via the Silk Road and be enjoyed by even more cultures. (en.wikipedia.org/wiki/Amber_Road)

With the demise of the Roman Empire by the early fifth century A.D., the Dark Ages set in and the road became less important. Literary references to amber and the route disappear, but the people still retained their love of the stone. During the medieval period the road was slowly reestablished, but never regained its importance as a major trade route.

The Amber Road has been credited with expanding trade between peoples whose paths may never have crossed, except for their desire to buy and sell this material. Discovering the path that amber took helps to understand how civilizations spread. Identifying amber by geological source allows scholars to follow it from its source to other locations, thus recreating the path that civilizations took.

Scandinavian countries especially prized amber and were part of the amber trade. Along with amber came European ideas that influenced their civilization, such as being credited with giving rise to the Nordic Bronze Age. This simple, organic stone had lasting impact on many civilizations.

The Amber Road was created to transport amber from where it was discovered to civilizations that prized it above gold. Most of the routes followed rivers and then overland to other rivers spreading civilization and allowing for cultural exchange.

SOME EARLY AMBER TRADE ROUTES

LEGEND

Northern Border Roman Empire — Eastern Amber Route
Amber Route Ancient Bronze Age — Sea Route
Main Amber Route Roman Era — Middle Bronze Age
Alati Amber Route Ancient Bronze Age

Artist rendering of Egyptian ear studs. When the Pharaoh screwed them onto his ears, only the face of the earrings protruded. Baltic amber and other resins were used to make these adornments.

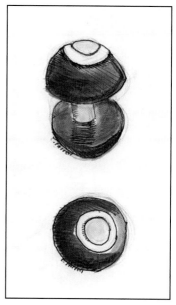

Amber Products

Amber has been commercially collected and mined since around 1264 A.D. Four types of amber products have developed: jewelry, smoking articles, objects of art, and devotional items. They all were made from collected and mined amber.

Jewelry

Jewelry was fashionable because the stones utilized in its creation were beautiful, easy to work, lightweight, and because they possessed powers that could enhance the person who wore them. A lot of myths and legends contain references to amber and contribute to its lore. Jewelry was worn to place the powers of the stone onto the wearer.

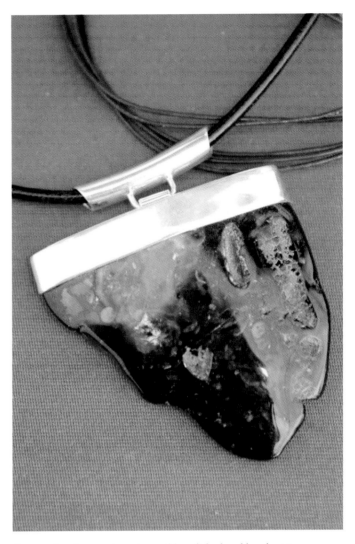

A natural amber specimen is roughly polished and bezel set in sterling to create this pendant. *Courtesy Amber Castino.*

Butterscotch amber adorns this money clip. *Courtesy Private collection.*

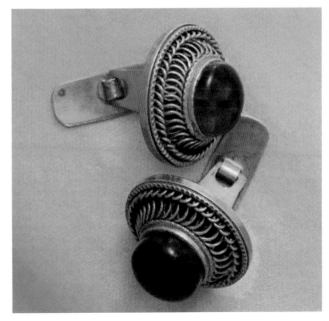

Cognac amber is the central stone in these ornate cuff links. *Courtesy Private collection.*

Smoking Articles

More than half of the amber harvested in the 1920s was used in the creation of smoking articles. This included cigar and cigarette holders as well as mouth pieces for pipes. Amber was believed to be an antiseptic. Smoking from an amber tip was considered safe, since it was believed to be sanitized. This belief encopuraged the sharing of pipes and other amber articles.

Amber cigarette holders were the passion of smokers before filters were added to cigarettes. This butter amber cigarette holder came in its own case attesting to the fact that it was natural amber. *Author's collection.*

Amber's antiseptic qualities made it perfect for smoking articles. This Meerschaum pipe sports an amber stem. *Author's collection*

Tisha is enjoying her amber cigarette holder as she smokes. The holder acts as a filter and keeps the nicotine away from her hand. These tips fell out of fashion when cigarette makers added filters to their product.

This honey amber cigarette holder was discovered in an antique store in Boone, North Carolina. *Author's collection.*

Red bakelite was often utilized in the creation of cigarette and cigar holders. *Author's collection.*

Amber cigar holders and cases were prized by gentlemen as a sign of status. *Courtesy Bubba and Jeannie Lesage, Clinton Collectibles.*

75

Objects of Art

Objects of art were carved from amber since early times. They appeared as jewelry boxes, cups and dishes, writing utensils, chess sets, and even chandeliers. A beautiful amber cabinet was created at Konigsberg, Germany, in 1743, and an amber ship was made there in 1934. The Amber Room in Russia (see page 135) was fashioned from slabs of amber that were fitted together and used as building materials.

Amber has been transformed into many religious objects. Paternosters and rosaries utilized amber beads and amber crosses. Alters were created from amber. Because it was as transparent as a window, amber was a desireable material for beautiful windows. Gold leaf was placed under the inlays so the amber shone from within.

Dylan is rubbing an amber ball against his wool topcoat to create static electricity which lifts the lint from his garment.

Amber has other uses. Skin-care products utilize the powers of amber to make women seem more beautiful and amber has been ascribed with medicinal properties. Balls of amber have been created and sold to remove lint from clothing. When rubbed on wool or other materials, amber produced static electricity that attracted the lint.

A modern amber rosary using white and Shelly amber beads combined with sterling silver provides a beautiful guide for remembering Jesus' story. *Courtesy The Sterling Rose jewelry.*

Amber balls were used to remove lint from clothing. The pressed amber, amberoid, specimen is an interesting sphere. *Author's collection.*

Pieces pieces of amber not fit for other purposes have been crushed and turned into varnish and shellac. These protective coatings from amber were created as early as 250 B.C., when they were used to protect and preserve works of art as well as furniture. The vivid, transparent colors associated with the Renaissance and Baroque masters of the fifteenth through seventeenth centuries can be attributed to painting mediums and varnishes that contain high concentrations of amber. Amber resins were known as Bernstein, Siccinum, Agetstein, and Glassa. When dry, amber varnish is impervious to most solvents and preserves the colors under the hard, yet flexible protective surface. (www.ambervarnish. com/index.php?pr=secrets)

Amber varnish was usually made in Germany, but other countries also created it. According to recipes discovered in old manuscripts, Baltic amber was melted under great heat and then cooled by dropping it onto iron plates. This produced a resin product that was splintery and soft. It still was not readily dissolved and needed to be heated again in an oil and cooked with litharge to a temperature of about 650 degrees Fahrenheit. This step incorporated the amber into the liquid. Some recipes mention that turpentine was added as the mixture cooled down. The heating of this flammable material produced a darkly colored varnish. Amber varnish utilized as a painting medium worked best with dark colors, but experienced painters knew that a mere drop of amber varnish gave hand-ground paint better binding ability, good paint character, and excellent paint durability. (www. jamescgroves.com/germanambervarnish.htm)

Musical instruments were also frequently coated with amber varnish. It dried in as little as one or two days without the need to place the instrument in direct sunlight (which could be damaging to the instrument). The coatings were built up in successive, thin layers that added an orange-brown coloration and glow and took a high polish. (www. jamescgroves.com/germanambervarnish.htm)

Powdered incense has been used for centuries to remove unpleasant odors. Amber is still a component of incense today. *Author's collection.*

Amber was utilized as a deodorant before room sprays were invented. Amber powder was used to enhance the smell of incense and served its place in rituals. The scent of burning amber was strong enough to cover the smells of rotting foods and unwashed bodies; thus has been a highly desirable commodity since before the medieval ages. (www. emporia.edu/earthsci/amber/uses.htm)

Amber essence is sitting atop a bakelite bead. Most of this essence contains little amber. Check your source to insure the product has the succinic acid which provides the best scent. *Author's collection.*

Amber resins made into varnish were used by the Masters to preserve their paintings. *The Alchemist* has re-created the formulas for today's artist. Dark and light varnish use linseed or walnut oil as the carrying agent. *Author's collection.*

Amber Lore

Where did amber originate?

There are many stories that try to explain amber's origin. Tears and water are prominent themes in these myths.

A Norse Myth

The ancient Norse people believed that amber was formed from the tears of goddess Freyja as she cried beside the sea. In their myth, she was the goddess of beauty, love, and fertility who was married to Odur, the sunshine of her life. She was a beautiful, blue-eyed blond who lived a charmed life, but she had a weakness for other things of beauty, especially jewels. Freyja and Odur lived happily in the land of Asgard in her palace, Falkvanger, with their two daughters. She was the queen of Aesir and had the means to purchase what she desired.

However, all was not beautiful in the land; the Black Dwarfs lived on the border of her kingdom. They were a crafty lot and equally great craftsmen. One day as Freyja was walking, she came across them as they were working on their craft. They had created a stunning necklace made of gold. It was named the Brisingamen or Brising necklace. It shone with the blinding brightness of the sun. She knew that she could not live without it.

Freyja begged the dwarfs to sell it to her as she had never seen anything to equal it before. All that she already possessed was forgotten. And when she was told that the piece was not for sale for all the silver in the world, she inquired, what treasure she could offer for its purchase.

The dwarfs told her that she could have the piece but that she must purchase it from each of them for the treasure of her love. Freyja was to wed each of them for a day and a night before the necklace would be hers. Madness overcame her and her desire for the necklace was so great she forgot her family and agreed to the unions. She remained with them for four days and nights. After these unholy unions, she was awarded the necklace and returned to her palace.

Her madness had now subsided and she felt great shame at what she had done. Upon entering the palace, she went to her chambers where she hid her costly treasure. Her misdeeds were soon discovered by Loki, the mischief maker, who went to Odur to inform him of her betrayal. Her husband did not believe that she could betray him and demanded proof.

Loki turned himself into a flea and flew into her bed chamber where she was sleeping. There on her neck was the necklace. Loki bit her on her cheek, causing her to turn her head so he could unclasp the necklace. He carefully took it from her neck and proceeded with it to Odur.

After seeing the necklace, Odur knew the truth, cast the necklace aside, and left the palace. He wandered into lands far away hoping to remove himself from her treachery.

Upon awakening, Freyja discovered the Brisingamen was gone. She knew she had to tell her husband the truth and sent for him. It was then she discovered the real price she paid for the jewel. Her true love was gone. Tearfully, she went to Valhalla to confess her misdeeds to Odin, the father god. As she entered Valhalla with Loki carrying the jewel, she passed through an amber grove called Glaeser that had trees dripping glistening amber.

Father Odin forgave her for her sins but took the necklace from Loki and placed it upon her neck as a reminder. He decreed that she must wear the necklace forever to remind her of her past misdeeds. She then set out to find her husband. She wandered the world crying over her loss. Her tear drops fell to the ground and where they landed on earth, they turned into gold, but when the landed in the sea, they turned into amber. (Rice, 1993, 115 – 116)

This tale reminds the Baltic people to tell the truth. It is believed in that region that an amber necklace chokes the wearer who forgets to tell the truth.

Artist rendering of Freyja, a Norse goddess, who lost her love because of infidelity and her lust for treasure. She grieved over her loss and cried amber tears.

A Greek Myth

The ancient Greek people also believed amber to be derived from tears. In the myth *Phaeton*, the story is told of a young man, Phaeton, who grew up not knowing his father, the Sun God, Helios. He asked his mother to give him proof of his ancestry and one day she agreed. She told him that his father was indeed a god and that he should go and question him. Phaeton set out to find the sun god to ask if the story was true.

Phaeton traveled to India where the sun god lived in a palace shimmering with gold and precious stones. Here he entered the hall and was blinded by the light. Helios dimmed his radiance when he saw his son enter the hall and asked Phaeton to approach. Phaeton asked if he was Helios' son and was told that he was. Phaeton was unable to accept the truth and asked a favor from his father as proof. Helios promised to grant him a wish so that he would know he was indeed his son .

The boy asked to drive the flaming chariot of the sun across the sky. His father offered other gifts, but Phaeton would not listen. Warnings were given about the dangers that could confront the would-be charioteer. Cautions were spoken to the skill needed to manage the chariot, but Phaeton would still not listen. Even though they both knew the dangers, Helios finally gave his consent.

The daughters of Helios, the Heliades, helped their half-brother to yoke their father's steeds on the day of the event. As dawn broke upon the morning sky, Phaeton leaped into the chariot, took the reins, and raced off to the west. Soon he left all behind as the chariot ascended into the sky.

The horses soon realized that Helios was not in his normal position and that an inexperience driver held the reins. They became wild and strayed from the usual path. Phaeton was paralyzed with fear, lost control, and even forgot the horses' names. He dropped the reins and became extremely dizzy. Sensing their total freedom, the horses raced wherever they pleased. When they came too close to the earth, the clouds began to smoke, harvests burned, the fields became parched, and the rivers began to dry up. Even the seas shriveled because of the heat.

The Earth Goddess asked Jupiter to help. He sent a thunderbolt at Phaeton causing him to fall into the Eridanus River where he died. His half-sisters were very distressed by his behavior and the destruction that he had caused. Because they had helped him and encouraged him in his destructive ride, they became rooted to the spot, changed into popular trees, and still to this day weep tears that harden and became amber. (Williams, 1938. 3-9)

The Heliades, sisters of Phaeton, are depicted in the artist rendering. Because of the part they played in his disastrous ride in the Sun's chariot, they are routed to the ground as trees and doomed to cry amber tears over his coffin.

The Sophocles Myth

In another story, Sophocles, who might have received a piece of amber with a feather inclusion, thought that amber was tears of special birds that cried at the death of Meleager. According to legend, Meleager's sisters were transformed into birds and left Greece once a year to travel to lands beyond India. As they flew, they cried for the loss of their brother. Their tears turned into amber and were spread over the lands where they traveled.

The Pliny Myth

Tears were not the only focus of these myths. Pliny, in the first century C.E., wrote that amber was the solidified urine of the lynx. Darker amber was from a male and lighter amber was from a female. He believed amber to be more valuable than a healthy slave.

Other Origin Myths

Another myth speaks of amber as the "juice" from the rays of the setting sun. (Conway, 1999, 36-37)

The early Chinese believed that the souls of tigers were transformed into amber at the end of their earthly existence. As you read the folk tales and mythology of the region, there are many other beautiful stories that attempt to unravel the mystery of this treasure.

No matter what the ancients believed about amber's origin, they respected it enough to attribute its special characteristics and considered it a desirable possession.

Amber Powers

Throughout the centuries, amber has been credited with many special powers:

Wear amber to attract good, loyal, and generous people into your life.

Carry amber to increase your logic and to allow you to use your wit in difficult situations.

Amber in the Stone Tarot suggests change or the "death of something". It represents renewal, beginnings, endings, health concerns, and trials that are being or will be faced.

Amber has been worn as a stone of prosperity to insure that efforts can at last be recognized.

Necklaces of amber were worn to protect the wearer from negative magic and this was especially effective in protecting children.

Jordan enjoys wearing an antique amber necklace and pin. She is unaware that the jewelry is there to protect her from negativity. She simply thinks they are beautiful. *Author's collection.*

Ashlyn places an amber necklace on her head as a crown. She looks like a princess of old. The necklace of older Baltic amber chunks is perfect. *Author's collection.*

Shamans wore amber beads to strengthen their spells. Wiccan High Priestesses (White Witches) wore alternating amber and jet beads in necklaces.

During the Renaissance when buxom females were in vogue, amber was worn because it was believed to increase body weight and make the woman more beautiful... *This should never be told today or amber would truly fall out of fashion.*

In ancient times, when sexual activity was considered normal, amber carvings in the shapes of male and female reproductive organs were popular. They were believed to have the power to increase sensuality and enjoyment of physical pleasures.

Amber and jet were worn by Wiccan priestesses because of the powers the combination was believed to possess. *Author's collection.*

Kim enjoys wearing amber and jet because they perfectly compliment to her dress. *Author's collection.*

81

Images of fish, frogs and rabbits were carved in amber and carried by women who wanted to conceive children. Men desiring to father children wore or carried amber images of lions, dogs and dragons to insure their success.

Because of the transparency and translucency of amber, it was believed to strengthen eyesight.

Amber is thought to be the stone favored by those born under the sign of Gemini, Leo, and Virgo.

Women desiring to get pregnant carried fish, frogs, and rabbits to improve their chances. This detailed carved amber fish looks as if he needs to return to the waters and spawn. *Courtesy The Amber Museum Gift Store, Santo Domingo, DR.*

Amber animals were carried or placed around people to transfer the magic the amber figurines held to the bearer. A clutch of turtles is beautifully sculpted in white amber. *Courtesy Amber Mundo Museum.*

Native craftsmen in the Dominican Republic still carve fish, rabbits and other animals from amber. *Courtesy Patricio Jhonson, DR.*

As a stone of the Sacral Chakra, amber is a good stone to improve memory, relieve stomach problems, and aid in digestion.

On the emotional level, amber is good at relieving depression and anxiety. It allows a person to recognize his/her self worth and value, thus feeling better about himself/herself and the world.

Amber is a stone of the intellect and helps to easily increase left/right brain integration.

This artist rendering of the Sacral Chakra represents an energy center that governs one of the lower abdominal areas and provides us with an emotional connection to the world. Amber is a stone used to heal wounds or balance issues in this chakra.

Because amber melts and appears to flow, it is an excellent stone in the workplace where it can help melt rigid and confrontational attitudes and blend them into a less competitive place.

Amber is recommended for shy or withdrawn children as it can aid in allowing them to relax in social situations. It can also protect them from the criticism that we all face in the classroom and on the playground.

The Romans utilized amber as an amulet which protected the gladiators from both danger and fear.

This beautiful orange amber specimen weighing about four pounds contains natural sediments from the matrix. The material is left natural until the carver determines how to best utilize the piece. *Courtesy Patricio Jhonson, DR.*

Amber is usually deposited in layers or flows. This red-toned Dominican Amber clump weighing over 2.5 pounds shows the irregularity amber takes from successive flows or drippings. *Courtesy Amber Mundo Museum, DR.*

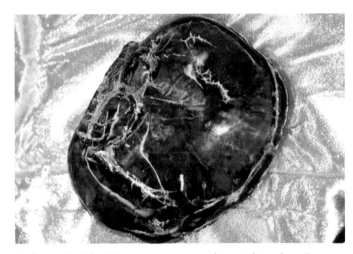

A clump of polished Dominican cognac amber weighing about 5 pounds shows some of the matrix left after polishing. Each crack shows the edges of the deposits. *Courtesy Turi Gift Shop, DR.*

Persons desiring astral travel hold amber up to the light and use the radiance of amber as a window or doorway to their travels on the amber path.

Looking through amber allows the gazer to enter a dream world. Vanessa is holding amber up to the light and dreaming about her future.

Amber comes from the seas and is believed to cleanse the waters from which it came as well as other polluted bodies of water.

Because amber has electrical powers it was believed to be an important stone for attracting love and lovers.

Attracting positive energy into your life and removing the negative energy is one of its most valuable powers.

Amber is a warm color associated with the sun and therefore it is a stone that can bring us warmth in times of need and it should be carried or worn by anyone recovering from surgery, illness, or injury.

Since amber is derived from a living thing, it is believed to carry a life force within it that can be used in times of need. This life force brings a form of energy into our lives than infuses our bodies with the strength to go on. This is especially helpful for the elderly and infirm.

Amber is a very rare stone. It was formed on the surface of the planet, not in the bowels of the earth like most other minerals. It was made by plants utilizing the sun's energy to grow and develop. Therefore it is a form of solidified sunlight and can bring this light into your life.

Amber was used to treat asthma and edema.

Persons suffering from hemorrhoids were told to fill a chamber pot with burning coals, place amber shavings on the coals, and to sit over the pot to relieve the condition..

Amber retrieved from the sea is placed there again to cleanse the water. Kenneth, Kaylie, and Hunter are placing amber into the Gulf of Mexico to allow the stone to cleanse it from pollutants.

A person suffering from hemorrhoids was told to scatter small pieces of amber over burning coals in a chamber pot and sit over them to relieve this condition.

Amber was reduced to a powder, mixed with honey and rose oil and used as a treatment for deafness.

Amber made into an elixir was believed to strengthen the thyroid. Amber beads were also worn at the throat to reduce goiter.

In China, amber syrup was used as a tranquilizer and to control muscle spasms. This syrup was made from succinct acid and opium.

Craftsmen who worked amber during the plagues did not die or even contract the diseases. Amber was believed to be an antidote.

Amber has been used in many medicines and when powered amber was mixed with honey it was believed to cure eye, ear, and throat diseases.

Stomach illnesses were believed to be cured by drinking water that had been steeped with amber pieces.

Eastern countries believed that amber smoke strengthened the human spirit and allowed the inhaler to capture courage.

With beliefs like these and others, it is easy to see why amber has held such a special place in the hearts and minds of civilizations over the centuries.

Amber Today

Carving and Crafting Amber

Care must be taken when carving or crafting amber. It is a soft stone with a low melting point. It can also be somewhat brittle and can fracture or shatter if it is not handled properly. Before working with a raw piece of amber, make sure it really is amber. Other materials will not react in the same manner.

Once it has been determined that the material is amber, craftsmen will check the piece for any cracks or areas that need to be avoided. Inclusions can increase the value of a piece and should be situated to the best advantage. Next the design is laid out.

Most amber is worked by hand with a jeweler's saw and fine toothed files. It can be worked with a flexible-shaft machine like a Fordham or Dremmel or polished with a polishing wheel, but craftsmen need to remember to use a very slow speed. Too much speed could cause friction and melting. Craft saws, jewelry files, and sandpaper can also be used. It is very important to always keep the flex-shaft moving and to move slowly but continuously. If the machine is stopped with a saw or bur in the amber, it may remain melted in the material ruining both the tool and the piece.

Pena, an artist at the Amber Factory in Santo Domingo, D.R., shows how he first removes the crust or skin from a piece of amber using a grinding wheel. *Courtesy the Amber Factory, Santo Domingo.*

When the shape of a design is laid out, any surface cracks and blemishes are removed with any of the tools. Sanding discs and large round burs can be used with the flex-shaft or with a grinding/polishing wheel to make this step easier. When sanding, craftsmen start with rough wet-sandpaper about 80-grit. Higher and higher grit sand paper is used to complete the piece and 600-grit paper is used to finish the piece. Amber carvers must always remember to have the piece and hands well braced on a workbench to insure control.

After removing the crust, Pena polishes the amber on a felt wheel. *Courtesy the Amber Museum, Santo Domingo, D.R.*

After cleaning and polishing the piece, the black amber is ready to use in jewelry. *Courtesy the AmberFactory, Santo Domingo.*

If a pendant is being crafted, how the piece will be hung must be determined. The piece can be set into a bezel, but care must be taken when bending the bezel over as the amber is soft and can be scratched or broken. Some craftsmen use pure silver or very high carat gold as the bezel material because it is softer than sterling or 14 K gold and can be worked much easier. The pendant can be wire-wrapped for an interesting look. Amber can also be set with prongs.

If the piece is going to be drilled, that decision needs to be made in a very early stage before time and energy is put into carving - just in case the amber shatters during drilling. Most amber is transparent and the cord or wire that is used will be visible in the completed piece.

Before drilling a hole through the piece, a line is drawn with a pencil at the level desired. The pieces are then looked at from the other two sides and similar lines are drawn. Where the lines intersect, the piece will be drilled. Artists make sure that the piece will hang as desired once it is drilled. They make a small indentation where the lines cross at the top and the bottom using a small round bur. Once this has been made, a small drill is used to create the hole. If the piece is short and drill is long enough, it can be drilled straight through. Most pieces are dry drilled from both ends and meet in the middle. Drilling is done slowly and steadily, using slight pressure. The drill is pulled out frequently and the amber dust is blown away. Close attention needs to be paid to where the drilling is done so that the holes meet in the middle. Craftsmen never stop the drill in the middle or the drill will get stuck. If the hole is off, a larger and larger bit can be used until the hole looks right. The hole will need to be thrummed before stringing the piece.

Now that the hole has been successfully drilled, carving continues on the piece with round burs. Close attention must be paid to the direction that the drill turns in order to spin it away from the edges to avoid chipping the piece. Once the design is complete, the final polish is added. Sanding the pieces with finer and finer sandpaper is done until it gives the look desired. Artists use 600+-grit sandpaper for the final sanding. After sanding, the piece is again polished using brass polish on a soft cloth.

Pieces that have been drilled need to have the holes cleaned and smoothed before stringing. The hole is thrummed with a piece of string that has been tied to the workbench and coated with olive oil or another oil. The process is repeated with other strings coated with silicon carbide powder, then a finer silicon carbide, next Tripoli (not oil), and finally brass polish. This allows the stringing material to slide much smoother and last longer.

After the final polish has been applied, the final piece is assembled the treasure has been created for our enjoyment. (www.lapidaryjournal.com/jj/899jj.cfm, *Carving Amber* by Yoli Rose.)

Because amber is soft, most amber is bezel set. The bezel protects the amber from scratches and breakage. The rings shown here are surrounded by sterling bezels. *Courtesy Marina Szing.*

The three dimensional effect of this pendant is achieved by carving intaglio. *Author's collection.*

A softer look is achieved by wire wrapping. The wire protects the edges and the curves while the wire wrapping catches the eye. *Courtesy The Sterling Rose jewelry.*

The same piece, when viewed from the back, shows the workmanship required to produce this unique pendant. A red dye has been applied to the back so that the carving could be more intense. *Author's collection.*

There are artisans in Poland and the Dominican Republic that practice the art of intaglio in amber. A piece of shelly amber, clear amber, is transformed into the basic shape desired for the piece and finished as described. Once the final polish has been added, the amber is carved. This time the carving is made into the amber from the back. When the piece is seen from the front, it appears to have something imbedded in the amber, but from the back, you can see the true genius of the carver. The design is reverse carved into the stone. As a final treatment, dye can be applied to the back to make the design clearer and to show a layer of color.

Other artisans create beautiful mosaic designs that show off the colors of amber. As in the Amber Room, gold leaf or other treatments can be applied to the back of the amber to bring out a beautiful pattern of color. This treatment is considered normal, acceptable, as it enhances the piece. The treatment does not need to be disclosed because it is considered permanent.

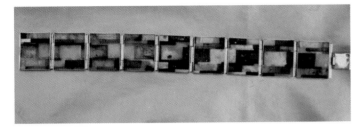

Mosaic work in amber is almost a dying tradition. There are few crafts-people left who take the time to create bracelets like this one. *Courtesy John McLeod.*

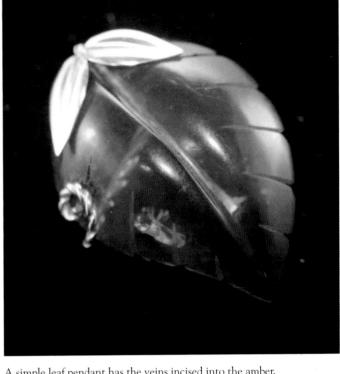

A simple leaf pendant has the veins incised into the amber. *Courtesy The Sterling Rose jewelry.*

Extremely talented artisans are creating beautiful carved pieces. Roses, leaves, and other floral design are being carved into amber and carry a hefty price. Animals are carved from amber for the enjoyment of the beholder. Because it is soft but fairly tough, amber has been treasured as a carving material for centuries.

Simple bands are left raised on this amber as the rest of the amber is carved away. This delicate procedure produces a piece worth setting in 14 k gold. *Courtesy Amber Mundo Museum Gift Store.*

Amber is carved into other forms. This amber cat purrs with the confidence that it was well made. *Author's collection.*

Testing

Putting all the more popular testing methods in one place should make it easier for you to find and use them. No one method that can easily be performed without a lab is a guarantee that your piece is natural amber. Use several different methods as needed to test your amber. Remember that imitations, simulants, can be natural resins or synthetic resins as well as modern materials. The piece you are testing could also be improved amber; doublets or triplets, layers of elements that may or may not contain natural amber; reconstructed or pressed amber made from small pieces fused together; or a combination.

When you see "genuine amber" beware. This name is meant to sell you an amber product, not natural amber. This genuine amber has been made out of small pieces of amber that have been melted together under high pressure. The material is normally very even in coloration and shape. Natural amber does not usually have this perfect an appearance. The name is a legal way to sell you an inferior product. There are excellent craftspeople that can find a perfect piece of amber and shape it perfectly, so do not discount their work; but do be cautious.

The first thing to do when questioning whether a piece is natural amber is to look at and hold the piece. Amber should not be cold to the touch as it does not readily transmit heat and cold. It should be light in weight when compared to other materials of similar size. Since amber is deposited in flows, there should be some indication that the amber has layers or flows of color. The color of natural amber is usually splotchy; it should not show obvious swirls. Most amber has small imperfections and inclusions, consider them as beauty marks that identify a piece as natural.

While looking at the piece, observe whether there are any mold lines or other indications that the piece in man-made. Natural amber is never molded. Genuine amber might have been molded and most plastics were. Since plastics were made for the masses, the mold lines were usually never removed.

Some amber connoisseurs can identify amber by taste. After a piece has been washed with a perfume free soap and rinsed, a touch to the tip of the tongue should give a hint as to whether a piece is amber or plastic. Amber has a pleasant piney taste. Plastic has an acrid taste and a bit of a bite. Please make sure the piece is clean before trying this. With a bit of practice, you might be able to taste the difference.

If you have a black light, you might want to shine it on the piece you are questioning. Most amber fluoresces and appears to glow. The most common colors observed under ultraviolet light are yellow, blue, green and orange. Since all pieces do not have this color change, it should be one of several testing methods used.

Your sense of smell can be used to distinguish amber from other materials. If you heat a piece of amber with a hairdryer, the piece will give off that piney smell. Bakelite will have a foul formaldehyde scent. Plastics will have different scents. Never heat a piece you believe to be celluloid as that material is very flammable.

Bakelite collectors use a pink semi-chrome polish to test for this. When a piece is rubbed with a soft white cloth, bakelite will turn the cloth orange. Orange indicates the piece could be bakelite and is not amber. A bakelite piece that has been polished will not react.

Michelle is heating a needle until it is red-hot.

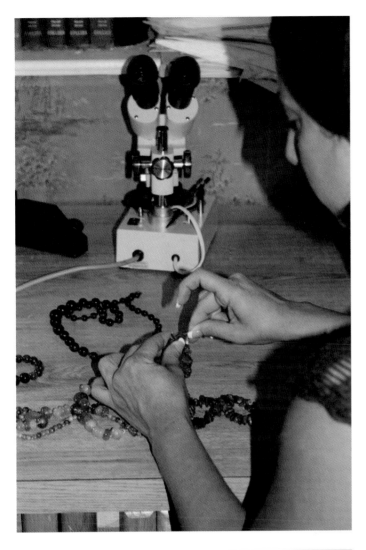

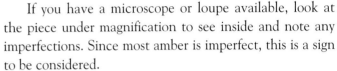
Once heated, she sticks the hot needle into the bead hole of an unidentified bead to see whether the material is amber. Amber has a piney scent, while plastic has an acrid scent. Care must be taken to touch the bead in an inconspicuous place so as not to damage the bead.

If you have a microscope or loupe available, look at the piece under magnification to see inside and note any imperfections. Since most amber is imperfect, this is a sign to be considered.

The rest of the tests are somewhat destructive and should be used as a last resort. The *hot point test* is a good way to test amber if you can get to an inconspicuous place for testing. A needle is heated to a red glow and pushed into the piece. Natural amber again has a piney or burnt resin smell. Fake amber has a more acrid smell. Copal will also give off a piney aroma and this makes the test somewhat inconclusive.

To test whether a piece is natural amber or copal, you can use the acetone or rubbing alcohol test. A small drop of acetone or rubbing alcohol is placed on the piece. After about twenty seconds, the copal will begin to melt; amber will remain unaffected. The surface of other simulants will also begin to melt. The acetone will leave a dull spot. NEVER test a conspicuous spot, since the dullness can not be removed and you could damage a collectible piece.

If you have a valuable piece, you can send it to a laboratory for testing. Infra-red spectroscopy can determine amber, as well as the type of amber.

Amber that has not been incorporated into jewelry can be tested by carefully placing the piece into saturated salt water. If the piece floats near the surface, it is lighter than the salt water and could be amber. Pieces that have been strung could be weighed down by the stringing material. Pieces with metal findings would be heavier because of the density of the findings.

Care

Because amber is a unique stone, special care should be taken with it. It has lasted for millions of years, but, it is fragile. If you follow a few precautions, you should be able to enjoy it for many years and even pass it on to future generations.

Remember that amber is brittle and special care should be taken when placing a piece anywhere. A sudden drop could cause the piece to fracture, ruining it. Carved pieces, amber containing inclusions, intaglios and mosaics should receive extra special care when being placed on a dressing table or other surface.

Amber necklaces hold up best when strung on silk or another soft wire and a knot is placed between each bead.

When putting amber away for storage, it should be protected. It should be wrapped, placed in protective pouches or stored in boxes to prevent it from being scratched or damaged.

Amber is soft and can be scratched with a fingernail. Store your amber pieces so they do not come in contact with sharp objects or harder materials. It is advisable to place pieces in individual pouches or boxes to prevent marring of the surfaces.

Amber necklaces should be strung so that a knot is between the beads. This will prevent the beads from abrading each other. (Many older necklaces are dated by wear in the bead hole. It will not take away from the value of the necklace, unless extreme). Amber necklaces should be hung when not worn to prevent them from tangling and give them a natural drape.

It is advisable to apply any hairspray or perfume and allow it to dry before putting on amber jewelry. If you frequently apply these after putting on your jewelry, a whitish film can be deposited on the jewelry which will eventually bond with the amber and become permanent. This dullness can devalue your pieces.

If you swim in a pool, please remove your amber jewelry before swimming. The chlorine can damage your jewelry as well.

Cleaning products, lard, salad oil and butter can damage your amber jewelry. It is best to remove rings and other jewelry that might come into contact with them before using these products.

Amber necklaces should be hung when not worn to prevent tangling. At the Turi Gift Store in Santo Domingo, Mariano Rodriguez, Hortencia Disla, and Mariano Jr. stand in front of their displays.

At the Coral Reef in New Port Richey, Florida, the Groves store their necklaces on shower rings and hang them on wire.

Ultrasonic and steam cleaners should not be used with amber jewelry. Amber can be cleaned by rinsing in clean lukewarm water. It can also be buffed with soft cotton or flannel.

Amber melts at a fairly low temperature. Do not keep it in an area that is subject to sunlight or high heat. Never leave amber in a car in the daytime. You may not like what you find when you return.

If you need to have repairs done to jewelry containing amber, the amber should be removed before any heat is applied to the piece.

Ultrasonic cleaners, like the one pictured, and steam cleaners should never be used on amber. The heat and vibration can damage the amber.

If jewelry repairs need to be done to a piece containing amber, the amber must be removed before any heating is done. This pendant has lost one bead and it can be glued without the need for torch work. *Author's collection.*

Most amber jewelry has been set in sterling silver. Since the 1990's, Polish sterling silver has contained anti-tarnish agents. This was mandated by law. Other countries do not have similar laws and jewelry created before 1990 or elsewhere in the world will not have these anti-tarnish agents. A lot of the amber jewelry on the market will need to have the sterling polished because it will tarnish. It is best to use a silver polishing cloth. The cloth is easy to use, does not leave a residue, and will not harm the amber.

Amber will naturally undergo a color change over the years. It tends to darken. Most red amber has oxidized over the centuries from lighter colored ambers to create its wonderful color. Oxidation can also cause crazing of the surface. Buffing and using olive oil can slow this process.

Amber has survived and been revered for millenniums. A little care on your part can insure that your treasures last as well. Wear your pieces and enjoy the warmth they add to your life.

If the amber lacks luster, it can be restored by dropping a small quantity of olive oil onto a soft piece of flannel and buffing the piece. Less is always best when using olive oil.

Amber that has lost its luster can be restored by gently rubbing it with a small amount of olive oil. Maranda is restoring the luster to some old Russian cuff links with olive oil.

A silver polishing cloth is the best tool for polishing the silver used to set most amber. Kimberly is polishing a piece belonging to the Sterling Rose.

Imitators & Enhancers

Because of the popularity of this stone, it has often been imitated. There have been many simulants that have posed as amber. These counterfeit stones do not have the same chemical composition as the real thing; do not have the same characteristics; and do not have the same value. There are also amber synthetics that are composed of amber, but are not natural. Many times they are treated, dyed, assembled, or enhanced. As already mentioned, the buyer must be very careful to only purchase amber from a reputable source. However, the synthetics and simulants are equally fascinating as long as you, the buyer, know what you are getting.

Many amber copycats contain inclusions of animals. There are two main things to remember: (1) Inclusions in natural amber are rarely perfect. Those sold as imitations usually have perfect inclusions and are sold as imitations by reputable dealers. (2) Vertebrates are rare in natural amber. Dominican amber has more than most other locations but even there, a reputable dealer is very important. Vertebrates and other interesting creatures such as scorpions and spiders are relatively common in counterfeits.

Let's look at some of these imitators:

Amberdan – A transparent yellow-brown plastic that resembles amber. It sometimes contains included insects or parts of them and is misrepresented as amber. – If Amberdan is touched with a hot needle, you will smell the acrid odor of plastic instead of the piney scent of amber.

Amberina – This was formerly a trademark, for a transparent art glass with colors ranging from light ruby red to pale amber. This glass has a higher luster, specific gravity, and greater hardness than amber. Amber has been carved into cups and other articles that might be fashioned as art glass, but the difference is easy to identify by luster and specific gravity as well as the cool touch that glass usually retains.

Amberoid or ambroid – A name given to pressed amber.

Amberina, a glass, has been mistaken for red amber cups and vases. This Amberina vase is much too large and evenly colored to be mistaken for amber. *Author's collection.*

Bakelite – A synthetic plastic created partially to simulate amber but it has a higher specific gravity and sinks in water.

Bernat – An amber colored plastic. It sometimes includes plant and/or insect matter. Bernat is manufactured in Germany. This plastic has a higher specific gravity and sinks in salt water.

This large, early Bakelite pin could be mistaken for cloudy amber. *Author's collection.*

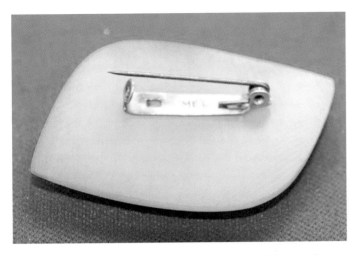

When viewing the back of the pin, the cutting and polishing marks are clearly visible. *Author's collection.*

Casein – The predominant phosphoprotein found in milk and cheese. It is used in the manufacture of plastics for knife handles and knitting needles. Again it has a higher SG and sinks in salt water.

Celluloid – The first thermoplastic and amber imitator was created from nitrocellulose and camphor in 1856. It was easily molded and shaped and became an ivory replacement, sometimes representing osseous amber as well. Knife handles, fountain pins, toys, and other items were fashioned in this material. Because celluloid decomposed easily and was flammable, its use was discontinued. Care needs to be taken when testing it with a hot needle. If it is rubbed briskly, a smell of camphor could be detected.

Copal (Persian Amber or African Amber)-An amber like resin that is usually softer than amber and much younger in geological age. Most copal is between 100 and 1,000,000 years old and still contains volatile terpines. Some forms of copal can be distinguished from amber by placing drops of acetone, alcohol, or chloroform on the material. The copal will dissolve leaving a dull spot. If you place a drop of ethyl alcohol on the copal it will start to soften within 20 – 30 seconds and if wiped with a cotton ball, the fibers of the cotton will stick. (www.crlc.ca/crlcart5.htm) Remember to treat an inconspicuous place to test the material. Copal from Columbia has begun to enter the market and its clear appearance closely resembles Baltic amber. Copal from Madagascar is also becoming available. Other copal can be translucent or opaque. Remember to ask before you buy and test if you doubt.

Cubic zirconia – in yellow and red has been utilized in Thailand to manufacture Buddha statues. The appearance should be a clue and the SG is high.

African copal beads are frequently sold as African amber. Most of this material is too young to be considered amber. *Courtesy Patricia Farrell.*

Old and new Bakelite beads are strung over the shoulders of Hunter. These large beads have been confused with African amber. The old beads are on the right and the new Bakelite is on the left. *Courtesy Joe and Eula Grove, the Coral Reef.*

97

James is testing Baltic amber and Madagascar copal. A drop of acetone is being placed on the amber.

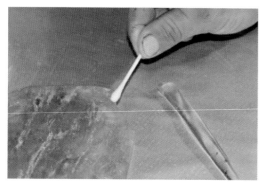

After a minute, the q-tip does not stick to the amber.

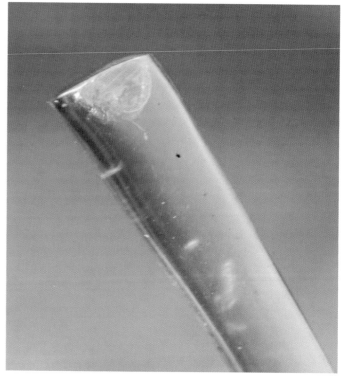

But the copal has melted and left a dull spot.

James is now applying a drop to the copal.

Dyed horn, birds' bills and other bones have been represented as Ossian amber. The appearance and texture are very different from amber and will also show porosity under magnification.

Antique glass beads also resemble amber. The coldness of the beads and the even color quickly let you know they are not amber. *Courtesy Marina Szing.*

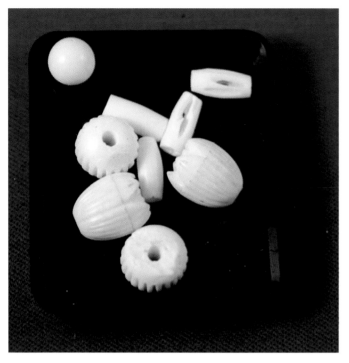

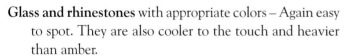

Bone beads can resemble osseous amber. Upon close inspection, these bone beads will show porosity that amber does not. *Author's collection.*

Glass and rhinestones with appropriate colors – Again easy to spot. They are also cooler to the touch and heavier than amber.

Kauri gum (containing embedded modern lizards) -- kauri is a natural resin from New Zealand or Australia and similar to copal. At the beginning of the 20th century, these synthetics with modern lizards were manufactured to fill a market need. They can be purchased for their antique value, but they are not amber. If you put a drop of alcohol on its surface, kauri will loose its transparency in less than a minute, amber will not be affected. Be careful to test in an inconspicuous spot or you could damage your specimen.

Lucite – A tough transparent polymer of methyl methyacrylate used as a substitute for glass and similar glass like products. Its SG is higher than amber.

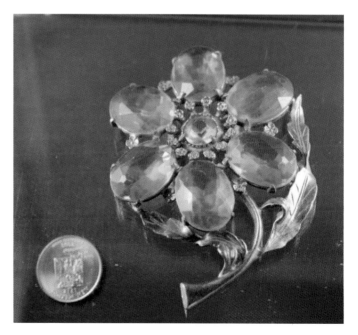

Rhinestones in this large pin are amber in color, but the coolness also gives them away as glass. *Author's collection.*

Another rhinestone pin. These pieces will be much heavier than amber and cool to the touch. *Author's collection.*

Plastic beads often are designed to imitate amber. These butterscotch beads are plastic. *Author's collection.*

This is yet another rhinestone pin. Because the color appears uneven, a new collector could be confused. Again, the piece will be cool and heavier than amber. *Author's collection.*

Plastics – A usually flexible synthetic material normally made of a hydrocarbon-based polymer. Bakelite is the best known plastic used to represent amber. Many times you will find mosquitoes, ants and ladybugs encapsulated in plastic made to resemble amber. Other times beads and bangles are made from bakelite with no inclusions. Bakelite is very collectible and jewelry made with it has increased in value. The SG again shows the difference as well as the smell when touched with a hot needle. Be careful Celluloid has also been used as a simulant and it is flammable!

Polybern – A polyester resin containing amber chips. This material has a brecciated appearance. You can see the chips of broken amber floating in the material. To tell if this is real amber you might need to observe it under magnification.

Prehnite – Massive yellow prehnite from Greenland can be mistaken for amber at first glance and so, it could be marketed as some kind amber. However it is harder and has a higher SG.

Pressed or reconstructed amber (also copal) – Amber chips and shavings are sometimes compressed together and sold as amber souvenirs. They are produced by applying pressure to relatively small fragments of clear scrap or to inferior amber or copal. They are fused in a vacuum with steam at about 400° F. The heat and pressure weld them together to produce larger pieces. Linseed oil or a clear plastic is sometimes added as a bonding agent either along with or in place of heating during application of the pressure. You should examine this material closely using transmitted light to see if it has color or transparency zones. Pressed amber often has elongate bubbles in the matrix and can be used as identifying characteristics. Sometimes the pressed amber is dyed, usually dark red. Inferior amber and amber scrap has also been used to produce amber rosin, amber oil and amber varnish, which is considered by some to be superior to synthetic varnishes.

Resins (colophony, gutta-percha, and sandarac) – Several of these resins that have reached the brittle state have been incorrectly called amber. A drop of ether, chloroform, or alcohol, when placed on most natural resins, will cause them to dissolve and a dull spot will appear.

Synthetic resins (catalin, cellon, celluloid, erinoid, galalith and rhodoid) - Each of these has a specific gravity greater than that of amber. They all tend to be easily cut as compared to the typical brittle quality of amber. Most resins emit an odor that is quite different from that of amber. Their odor is unpleasant, possibly rancid smelling where amber tends to have a sweet piney aroma.

Real butterscotch amber beads have a more polished appearance. *Author's collection.*

A detail of the plastic beads shows mold marks and distinct color bands.

101

A detail of real butterscotch shows layers of color and no mold lines or marks.

The real amber has a slightly different feel and look.

Plastic butter amber colored beads are light and not cold, but the uniformity of the color and look tells the story of the simulant. *Author's collection.*

Detail of the plastic shows swirls of color.

Detail of the amber shows layering.

These fan shaped earrings are plastic and worth a fraction of natural amber earrings.

Designers frequently imitate stones and designs that are popular. This Trifari necklace is collectible, but not because it's amber. *Author's collection.*

These "cherry amber" beads were bought from very reputable antique dealers. However, they are both bakelite. *Author's collection.*

A detail shows the uniformity of color that makes them questionable as amber and there are no inclusions.

Lucite was developed as an amber substitute. Buttons were frequently made of Lucite during the last century. *Author's collection.*

Notice the distinct swirls in this plastic bracelet. The colors are close to amber tones, but are somewhat off. *Courtesy Marina Szing.*

In a side by side comparison, a simulated amber bracelet and a natural amber bracelet are easy to compare. The bracelet on the left is not natural amber. It has been designed to resemble treated amber filled with sun spangles, but they have been overdone. *Author's collection.* The bracelet on the right shows characteristics to look for in natural amber. There are natural blemishes. The color is layered and the beads show quality workmanship. *Courtesy Gayle Robinson.*

Frauds

There also have been deceitful transactions. If you shop the internet, be aware of the following and other similar frauds.

A Baltic amber cabochon was available for sale that appeared to have an included insect. Instead, an insect was carved on its base.

A bead was produced in the 1980s by the Japanese ojime-maker, Tomizo Saratani, which looks like amber with included carpenter ants. The piece was made of ant-shaped forms created from black lacquer that were placed above a clear lacquer-coated amber core. This bead, with the unbelievably natural-appearing "ants," was unique. It is collectible, but it is not included amber.

An "amber insect pin," was offered for sale in a 2002 catalog of a reputable marketer. It consisted of a green amber cabochon set in a silver pin shaped like an insect. Do you think the purchaser thought s/he was purchasing an amber piece with an inclusion or a piece in the shape of an insect? Beware!

Be very cautious if you see an offer for an amber pin and no picture is available. Amber lizard pins and amber spider pins have been advertised in such a way as to make the reader believe that they have inclusions. However, you might receive a fun pin like the lizard pin pictured here. *Author's collection.*

As can be seen from the exmples above, the faking of amber has been around for a long time. It was most prevalent in the early 1900s when amber was desired for mourning jewelry. When England's Prince Albert died in 1861, Queen Victoria was in mourning for fifty years and acceptable jewelry included amber and jet. Supply was not equal to demand and synthetics were created to meet the desires of the British people.

New Zealand was especially successful at creating imitations. The North Island had major deposits of Kaori Gum and used it to produce the fakes. Business was so good that the Kaori workers had their own newspaper, "The Gum Diggers Gazette". After the gum was dug, it would be cleaned, melted gently and inclusions of colorful insects would be added. This color should be a giveaway to the item being a fake. Real inclusions have no pigmentation remaining.

This great amber spider is fun to wear, but not an included spider. *Author's collection.*

Some of the best fakes have involved real amber and ancient insects. The amber in these fakes was cut into two or more pieces and a hole was drilled allowing a small space for a genuinely old insect to be placed. Molten amber was used to fill the hole. Because of ambers low melting point this was fairly easy to accomplish. This forgery is hard to detect and a scientist who knows plant and animal life from the Tertiary or Jurassic Period should verify the legitimacy of the piece.

During the Victorian age, amber was thought to be better if it was clear and it was frequently heated just enough to remove the air bubbles and to clarify the amber. Some of it was also pressed giving it a more whitish appearance. Both of these techniques to enhance amber are not considered proper for "real" amber today.

Bees are sometimes found in amber, but this bee is made from amber and sterling silver. He has landed on a new bakelite bead that resembles amber. *Author's collection.*

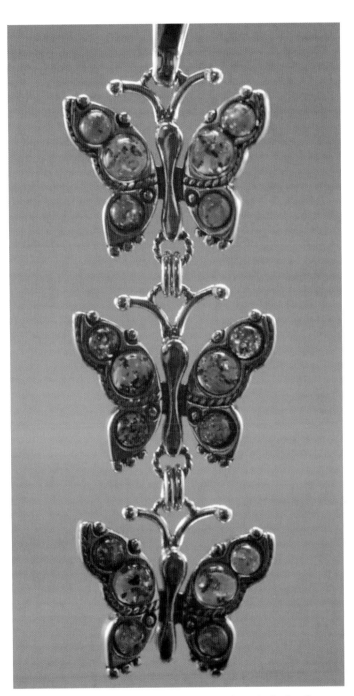

This trio of amber butterflies makes a nice pendant. *Author's collection*

A large butterfly is captured in pine resin. The butterfly is much too perfect to be the real thing. *Courtesy Joe and Eula Grove, the Coral Reef.*

Recent inclusions in large amber clumps are sold as reconstituted amber. This piece from the *Coral Reef* is beautiful and expensive, but not natural. *Courtesy Joe and Eula Grave, the Coral Reef.*

There is a technique currently used to clarify amber that is accepted practice today. Cloudy amber is placed into rapeseed oil and slowly heated in an autoclave. The bubbles found in the natural amber are filled with the oil and the amber looks clear. However, discoidal fissures or sun spangles are formed. When the spangles have brown edges, you know that the amber has been clarified. Some natural amber has these spangles from being warmed, but the brownish edges are a giveaway of treatment. This treatment does not need to be disclosed to buyers as it is considered to be permanent. The spangles are loved by American buyers, but most Europeans prefer their amber untreated. Currently, the Baltic area is where most amber is treated. Dominican amber rarely needs any treatment.

A lot of the green amber you see today has been clarified and then had the back either coated or burned to produce the green color. Both are acceptable treatments as they are considered permanent enhancements. *Author's collection.*

When cloudy amber is clarified by heating, sun spangles form. This is a pendant with treated amber. Compare it to the synthetic amber. *Courtesy The Sterling Rose jewelry.*

This lovely bracelet is a perfect example of what to look for in a quality piece. The cabochons are naturally shaped. The color is in the same family but different shades and tones are present. There are a few spangles to indicate treatment, but not too many. The silverwork is very well done. This piece commands a good price. *Courtesy John McLeod.*

Frequently there are only a few spangles in treated amber. You can see the brown edges in the amber set into this pendant. *Courtesy The Sterling Rose jewelry.*

Controversy

About 90% of the world's amber comes from the Kaliningrad, Russia, area. Drops or blocks of amber are mined there today from open pits. These pits contain glauconite sand, a hydrated potassium-iron silicate material called "blue earth" because of its blue-green color.

The amber is removed from the blue earth by clearing away the surface and then bringing in steam shovels and dredges to dig up the material. Conveyors move the earth to waiting freight cars and it is transported to spray houses. After the earth is transported, it is poured through grates at a washing plant where pressurized water separates the amber from the sand.

The first dredging began in 1854 and the excavation reached 35 feet below sea level. This process greatly increased amber recovery but began a degradation of the area. On one day in 1862, workers removed 4,400 pounds of amber. By 1865, the mining companies' 1000 employees were operating 22 steam barges and in a three year period had collected 185,000 pounds of amber. In 1875, the first open pit mine was created, and a million pounds of amber was removed in 1895. By 1930 the mine was fully mechanized and operations have remained almost the same since that time.

Damage to the environment occurs at the mines by removing a great quantity of earth and it also occurs at the spray houses where the run off is allowed to return to the sea. The Baltic Sea has been rapidly deteriorating over the last 50 years as more than 100 million tons of waste have been discharged there by the mining operations. The mining and processing produces material than can not be easily dissolved by the brackish water. This discharge impedes the natural flow of water in the Sea. Since the Baltic is cut off from the Atlantic Ocean, it can not easily refresh itself. Many environmentalists consider the mine to be one of the world's biggest problems when it comes to anthropogenic suspended material expulsion.

Another concern haunts the amber mines in this region, crime. The amber mines are huge and poorly patrolled. Pay is minimal and workers as well as treasure hunters can make more by illegally fishing for pieces that have been discharged from the mine through faulty disposal systems. Underpaid administrators sometimes look the other way as these gatherers share their profits with them. A large piece of amber can bring as much as four times the monthly salary of workers in the plant. Most administrators are unconcerned about the loss of revenue from this gathering as they are more concerned with large-scale theft and mafia involvement. According to a study made in 1996, annual losses at the Kaliningrad Oblast because of illegal gathering and smuggling amount to over $1 billion. (gurukul.ucc.american.edu/ted/amber.htm)

Gdansk, Poland

Amber has been treasured around the world and over the centuries, but there is nowhere that amber is more loved than in Gdansk, Poland. The earliest exhibits in museums attesting to the city's involvement with amber date it to the period 8 to 4 thousand years B.C. At that time, the locals produced beautiful amulets and other articles that made their way on the Amber Road. The earliest evidence of amber workshops dates to the late 10th century. Gdansk has had involvement in the amber trade for a long time.

Amber dealers who sell Baltic amber can be certified by the Amber Association of Poland. This certification asserts that they only sell genuine Baltic amber. Only treatments approved by the Association can be used in these products.

Today, people in Poland still have a very high regard for the quality of their amber products, and have established the Amber Association of Poland. This body offers certification that only genuine Baltic amber and treated sterling silver are used in jewelry sold by companies they certify. These companies must meet stringent requirements to qualify and to display the certificate. Buyers can be assured that the products sold by them are genuine.

Gdansk is now a destination in the tourist trade associated with amber. They hold the world's largest amber fairs, Amberif and Ambermart. Amberif, held in early March since 1994, is the world's largest showcase for amber and amber products. More than 6500 jewelry professional, gemologists, and scientists from around the world meet with almost 500 exhibitors to view, purchase, and enjoy the beauty of this treasure. In 2007, it was opened to the public for the first time. Other gemological treasures are permitted at the showcase, but amber is the star. It is interesting that amber jewelry at the fair is now found set with small diamonds, gemstones, and high carat gold, platinum, titanium, crystal, and pearls, as well as leather and silk. Amber is going high-end.

One opportunity for showcasing the desirability of amber jewelry is the Fashion Gala held at the fair. Each year, jewelry designers and Polish fashion designers present runway shows that demonstrate the versatility of amber in modern fashion trends. The show reveals that amber is no longer "grandmother's jewelry."

Scientists at Amberif are never disappointed; seminars are held in buildings specifically designed for their studies. There is ample supply or raw amber for their studies. (www.jcjonline.com/index.asp?layout=articlePrint&articleID=CA6452101)

In 2000, the Gdansk History Museum started a new branch as the Amber Museum, where you will find one of the largest amber collections in the world. Other museums in the city also feature amber exhibits and exhibitions.

The Museum of Amber Inclusions is located near the University of Gdansk, which has the only Chair of Amber. St. Bridget's Church features a monumental amber altar. Amber workshops, festivals, and galleries are a way of life there. (gdansk.pl/en/article.php?category=453&article=924&history=)

Lucjen Myrta, one of the great amber masters living today, has created amazing works of art with amber. An entire museum is dedicated to him and the treasures he owns and has created.

Poland is not the only country that treasures amber. Museums around the world showcase this amazing stone. Amber jewelry is worn with casual as well as formal attire. Objects made from amber grace shelves in homes and offices around the world. Searching the internet always brings up interesting information about the Gold of the North.

What other gemstone has captured the imagination and played such an important role in history. We wear amber jewelry; have traveled the Amber Road; admire the Amber Room; read books about the mystery of the missing amber; watch movies that showcase amber as a preserver of antiquity; utilize amber to cure our conditions and ward off evil; as well as create beautiful works of art with it. What more can a simple stone do for us?

Lucjan Myrta is the best known amber artisan living in Poland today. He has assembled an enormous collection of amber and has given it to Gdansk for an Amber Museum. He is now creating one of the most difficult amber creations ever conceived, a jewelry casket weighing over 500 kg. This artist rendering depicts what the chest should look like when completed. A true masterpiece!

Valuation

The popularity of amber has undergone fluctuation over the centuries that man has been fascinated by this unique gem material. In ancient times it was extremely valuable and has even been credited with changing civilizations. During the Victorian period it was again popular and expensive. In fact, simulants were made and eventually became more popular than amber itself.

Today, amber is again looked at as an interesting jewelry material. Ever since *Jurassic Park* ®, the movie about dinosaurs that were created from the DNA found in amber, was released, amber has made a comeback. Amber with inclusions commands the highest prices, while rare colors and excellent craftsmanship also contribute to the value of amber jewelry and objects. The size of the stone must be considered when reflecting on its price. Because amber is a desirable material, there are many substitutes available on the market, and a genuine piece of amber has more value than pressed or simulated amber.

Fran Wilson of *Amber Forever* created this amber and turquoise necklace for *Nieman Markus*. Due to health concerns, she no longer designs but her pieces remain spectacular. *Courtesy Fred and Carol Lamb, Payson Galleries.*

Amber was worth more than ten times the value of gold in ancient times. It does not have the same value today, but very expensive pieces are still there for your enjoyment, if you can afford them. This extensive nativity set is available for just over six thousand dollars. *Courtesy Amber Mundo Museum Gift Shop.*

A stunning saxophone pin crafted from sterling silver sports an amber bowl. *Author's collection.*

This complete Latvian amber necklace is very desirable and beautiful. Many shades and tones of amber are represented. *Author's collection*.

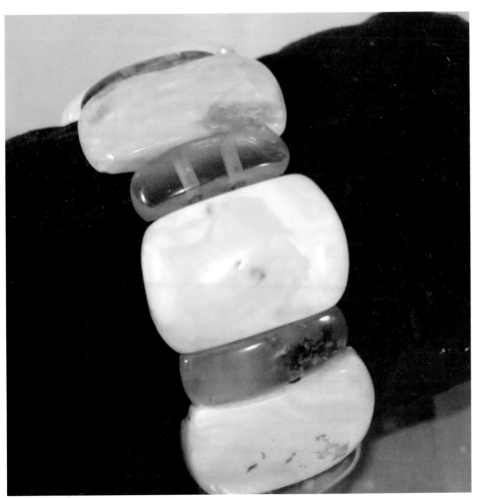

Bangles and bracelets are a great way to enjoy amber. This butterscotch and cognac piece is exceptional. *Courtesy Gayle Robinson*.

113

A matching necklace will frame the face of the wearer in its golden glow. Trapezoid pieces of natural amber are fitted together with slender spacers to form this amazing piece. *Courtesy Gayle Robinson*.

A four-inch flower with amber petals makes a statement whenever it's worn. This piece looks heavy, but the amber makes it lightweight and easy to wear. *Courtesy Clay and Cindy Oliver, Oliver's Twist Antiques.*

Cruise ships frequently showcase jewelry. A cruise from Florida to Mexico showcased amber. The inlaid piece shown here has a center section that rotates to expose another side and even more amber. *Author's collection.*

Christina is wearing an *Amber Forever* necklace with a intaglio carving on the pendant. *Author's collection*

A detail of the pendant shows a mythical unicorn sea horse carved into the amber.

During a telephone conversation (January 18, 2008) with Richard B. Drucker, GG, the publisher and editor of *The Guide,* a pricing guide utilized by professional appraisers to appraise jewelry, he stated that amber has been removed from the guide in the past year because of difficulty in valuing it. The number of fakes and simulants and the range of the market from antique to modern have made a simple formula impossible. Buying from a reputable dealer as well as knowing what you are buying is extremely important.

The current retail range for amber cabochons or chunks suggested by wholesalers is between $3 and $6 per gram for regular stock. The workmanship needed to create a piece of jewelry as well as the cost of the materials must then be factored in to the price of a finished piece. With gold in January of 2008 costing over $900 an ounce, and silver around $16 an ounce, these materials greatly affect the values of amber jewelry and other items created today, as well as the value of older pieces. The recent devaluation of the American dollar has also caused the value of amber in America to escalate. When amber has been faceted, cut into intricate shapes, or cut as an intaglio, its value is much higher. Rare colors also command higher prices.

Dominican amber specimens containing very desirable inclusions are now priced from $200 to $40,000. (www.gemstone.org/gem-by-gem/english/amber.html) Specimens only a few inches long, found in the Dominican Republic, and containing multiple inclusions now sell for $500 to $2500 and more.

Jewelry appraisers utilize *The Guide,* edited by Richard Drucker, to determine the value of gems and jewelry. The guide is now available on line for professionals.

A stunning array of amber and Larimar are available at the *Amber Factory* in Santo Domingo. *Courtesy the Amber Factory.*

Renninger's Antique Show brought an amber booth to Florida. John McLeod is proud that he is a certified amber dealer and only sells Genuine Baltic Amber. *Courtesy John McLeod.*

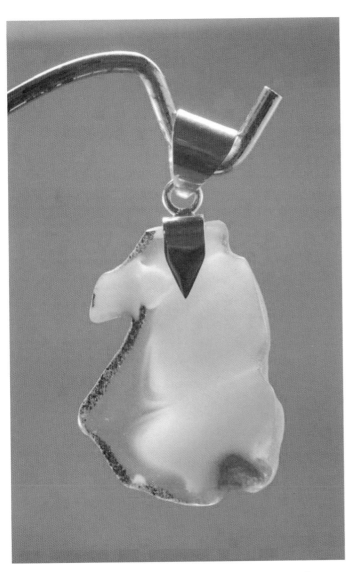

This lovely little pin has a vermeil setting. *Courtesy The Sterling Rose jewelry.*

Amber jewelry is priced for every budget from thousands to under twenty dollars. This unique amber slab is at the reasonable end. *Courtesy The Sterling Rose jewelry.*

Amber flip rings are an interesting addition to the wardrobe. Wear it as green amber or…

…flip the amber over and wear a cognac ring. FUN. *Courtesy The Sterling Rose jewelry.*

Design and stones add or detract value from a piece. This pin has lovely workmanship and design. *Courtesy The Sterling Rose jewelry.*

Amber bracelets set in sterling and gold are easy to wear and useful accessories.
Courtesy The Sterling Rose jewelry.

Some costume jewelry uses what was once considered semi-precious stones and pearls in their design. This pin has red and green died agate as well as amber and pearls. *Author's collection*.

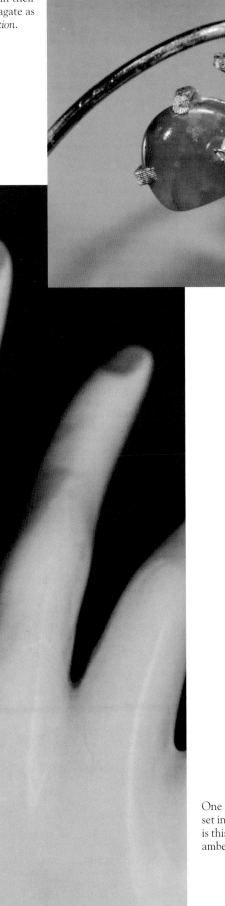

One would believe that amber would be best set in gold, but most amber is set in sterling as is this ring. The design sets off the glow of the amber. *Courtesy The Sterling Rose jewelry*.

Playful nymphs or fairies are designed with amber wings in these fun earrings. They are positioned against modern bakelite beads. *Courtesy The Sterling Rose jewelry.*

Dominican amber beads in two tones make a beautiful bracelet. *Courtesy Turi Gift Store.*

There are many internet sites and companies selling amber jewelry to the consumer. Prices on the internet vary with the quality as well as other factors, such as color, rarity, craftsmanship, size, number and type of inclusions, whether the stone is faceted or cut in a cabochon, and its place of origin. When shopping on the internet, be very cautious. There are products marketed with amber in their names that do not contain any amber. There are other sites that sell "natural" amber that is anything but natural. Most of this is created from bits and pieces of amber that have been heated and compressed into artificial forms of amber known as amberoid or Polybern.

This beautiful necklace and earrings was sold as a set for under forty dollars. Real amber would sell for several hundred dollars or even more. The spangles are beautiful, but not amber. *Author's collection.*

Looking at lovely natural and treated amber makes one wonder how the imitation looks. This sparkling amber bracelet was sold on the internet as Delightful Amber. When the buyer questioned what it was, the seller stated that its composition was unknown. The price also suggested that it was not amber. *Author's collection.*

This pendant looks like the real thing. The "stone" is not amber and the metal is not silver. Again fun to wear, but you get what you pay for.

Speaking about getting what you pay for... This beautiful cameo is an amber intaglio. The workmanship is remarkable. *Courtesy John Mc Leod.*

Russian amber cufflinks from the early twentieth century were worn with pride. Now they are interesting reminders of a time when shirt cuffs didn't have buttons. *Author's collection.*

A necklace of hundreds of amber beads strung together with amber chunk accents is a show stopper. *Courtesy Turi Gift Shop.*

Another pair of Russian cuff links have amber centers. *Author's collection.*

A butter amber pin with dangling amber drops from Czechoslovakia is a memoir of family members that are no longer with us. *Courtesy Marina Szing.*

Beautifully faceted amber beads surround an amber cabochon in this early twentieth century pin. *Author's collection.*

These faceted cherry beads are Bakelite, not amber, but are easy to wear and also very lovely. *Author's collection.*

Kaylie is wearing a faceted Russian amber necklace spaced with 14 K gold beads. The beads are perfect for formal or casual wear. *Author's collection.*

Large and small pendants alike compliment wardrobes. This musical pendant adds a note of inspiration. *Courtesy The Sterling Rose jewelry.*

Floral embellishments add to the beauty of this honey amber pendant. *Courtesy The Sterling Rose jewelry.*

Amber has many faces and colors. This unique pendant can be worn with either face out. It does dual duty. *Courtesy The Sterling Rose jewelry.*

Not all amber is used in the production of jewelry. At the *Amber Mundo Museum* in Santo Domingo, the banister to the second floor is made from amberoid, compressed amber. *Courtesy Amber Mundo Museum.*

A closer look at the railing reflects the beauty of the amberoid in the Dominican Republic.

A simple pendant can retail for as little as $8 to over $250 and a genuine amber necklace can wholesale for several hundred to thousands of dollars. Antique stores selling amber pieces will sell items in the same price ranges. Cherry amber necklaces will normally sell in the $250 to $350 range, but most of these necklaces are not amber; they are usually Bakelite. Bakelite now commands high prices so whether amber or Bakelite, the price is about the same. Necklaces made from chips do not normally command the prices that shaped beads bring. Most jewelry items manufactured from amber are set in sterling silver, keeping the price low. Some jewelry is now set in gold and commands much higher prices.

During the early years of amber's popularity, it was used to create religious jewelry. It still serves that purpose today. This inlaid amber cross reflects the warmth and love of God to the wearer. *Courtesy The Sterling Rose jewelry.*

The inclusions in amber make each piece unique. A simple oval cabochon and flat sterling bail create a gazing pendant. *Courtesy The Sterling Rose jewelry.*

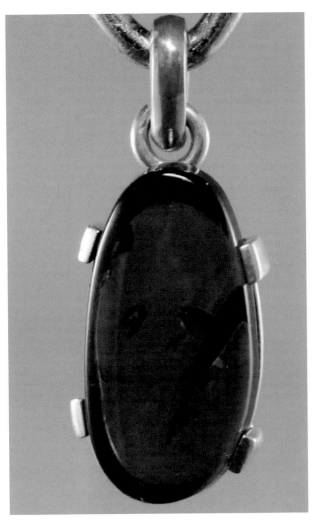

The classic oval shape sets off this modern pendant.
Courtesy The Sterling Rose jewelry.

A thick amber cross is a good reminder of faith.
Courtesy The Sterling Rose jewelry.

Chip necklaces have been a staple of amber collections The Baltic amber chips and Bakelite closing are typical examples. *Courtesy The Sterling Rose jewelry.*

Indian amber in dark and light brown sparkle in the light. *Author's collection.*

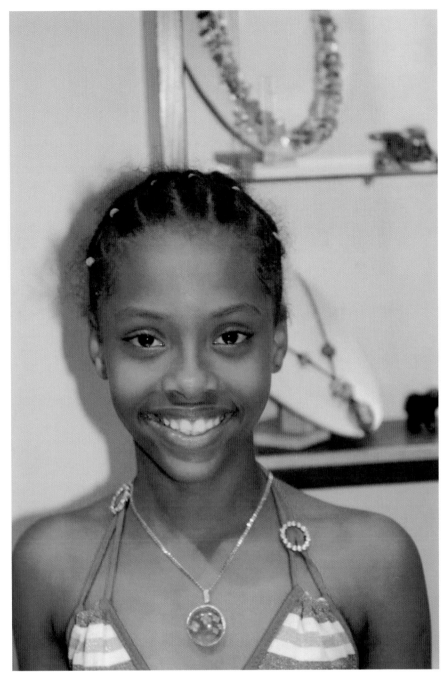

Kotiusa Montera Pacheco models a beautiful amber intaglio pendant set in14 k gold at her aunt's jewelry store in Santo Domingo. *Courtesy D'Marliyn Joyeria y Algo Mas, Santo Domingo, Dominican Republic.*

A closer look shows a lovely flower intaglio carved into the back of the piece.

133

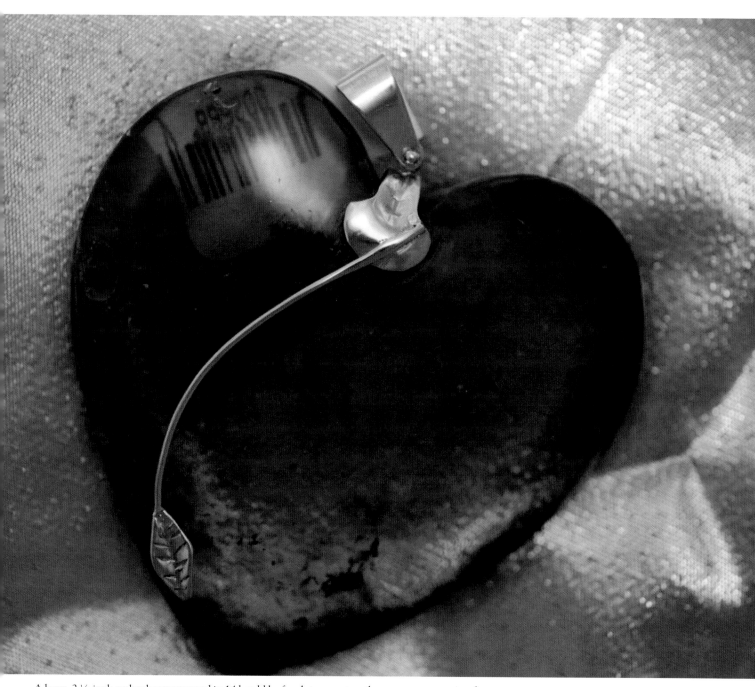

A large, 2 ½-inch amber heart wrapped in 14 k gold leaf and stem portray the wearer as a person of status.
Courtesy Amber Mundo Museum Gift Shop, Santo Domingo, Dominican Republic.

Special Projects Involving Amber

The Amber Room

One of the most beautiful creations utilizing amber is the Amber Room. This Eighth Wonder of the World was created between 1701 and 1709 in the Charlottenburg Palace, Prussia, the home of Friedrich I, King of Prussia. It was designed and created for his second wife, Sophie Charlotte. One complete chamber in the palace was decorated with amber panels that had been backed with gold leaf and mirrors. The room was designed by Andreas Schulter. The master craftsman on the room, Gottfried Wolfram, was also the master craftsman for the Danish court of the day. He had assistance from two other amber masters from Danzig (Gdansk): Ernst Schacht and Gottfried Turau.

An ally, Tsar Peter the Great of the Russian Empire, came to visit and greatly admired the chamber. The heir to throne, Friedrich Wilhelm I, presented the room to him as a gift to secure an alliance against Sweden in 1716.

In 1755 the room was transferred to the Russian Winter Palace and later to the Catherine palace by Tsarina Elizabeth. At this time more amber was needed to complete the design and the Baltic amber was sent to her by Fredrick II the Great of Berlin. The tsarina's court architect, Bartolomeo Rastrelli, from Italy created the new design.

The Amber Room underwent several renovations in the 18th century and after completion covered more that 55 square meters and had utilized over six tons of amber in its construction. This joint Russian and German venture was truly magnificent.

During World War II, the German army invaded Russia and Russian curators responsible for the Amber Room tried to protect it from being seized. They first tried to remove the panels to store them in a safe place, but the amber had dried out over the years and become brittle. When they attempted to remove it, it started to crumble. Not knowing what else to do, they covered it with wall paper in an attempt to conceal the treasure. However, the room was well known and their efforts failed. German soldiers discovered and disassembled the room in a mere thirty-six hours, crated it, and had it ready to be sent to the Konigsberg Castle in Germany.

Orders had been given by Adolf Hitler on January 21 and 24, 1945, that allowed treasures to be removed from oc-cupied territory by Albert Speer. Many treasures were taken by Nazi leaders for personal gain as well as to preserve them from destruction. The Germans were especially interested in the Amber Room because of its Germanic origin.

Erich Koch was responsible for the cultural items that were sent to Konigsberg, but he was not very successful in protecting them. Near the end of the war, the Germans tried to save the treasures they had confiscated and moved them to more secure storage facilities for safe keeping. Koch tried to protect the amber room but failed. With his failures, the real mystery of the Amber Room begins. Where does this treasure reside today?

There were eyewitnesses who saw the amber crates at the railway station in Konigsberg and believed that they were sent to the cruise ship *Wilhelm Gustloff*, that had left the area on January 30th, 1945. The ship was carrying about 9,400 refugees who had been trapped by the Red Army in East Prussia. The ship was hit by three torpedoes from a Soviet submarine while traversing the Baltic Sea. When it sank, it settled to the sea bottom in about 150 feet of freezing water. Almost everyone died and all the cargo was lost. The wreckage has been designated a war memorial and is off limits to salvage operations. We may never know if the crates made it onto the ship or what the crates truly contained.

Some believe the crates were stored in caves in Weimar and were never rediscovered. Many treasures were found in these caves, but many of them had been sealed with dynamite and covered with debris. All the hidden treasures were not re-discovered, and many may still be buried there.

Still other people believe the Amber Room remained in the Konigsberg Castle, assembled in a chamber for the Germans to enjoy. Some believe it was never crated, but was destroyed when the castle was bombed by the Royal Air Force and burned by the Russians. No sign of the Amber Room was found when the castle fell on April 9, 1945.

Currently, neither the Amber Room nor its remains have been located. Its fate remains a mystery, although there have been extensive searches made by the Soviet Union, the German government, and treasure hunters from around the world. Cries of outrage from the old Soviet Union contend their treasures were never returned and numerous conspiracy theories exist about their mysterious disappearance.

British investigative reporters, Catherine Scott-Clark and Adrian Levy, did extensive research on the Amber Room, including archival research in Russia and they wrote a book, *The Amber Room: The Fate of the World's Greatest Lost Treasure*, in which they concluded that the treasure was most likely destroyed when the Konigsberg Castle was burned shortly after it was surrendered to the Russians. However, officially, Russia continues to object to this conclusion, as they refuse to believe that their soldiers could have destroyed the treasure.

An interesting amber room novel was published in 2004. *The Amber Room* by Steve Berry is a fictionalized account of what might have happened to the Amber Room if it disappeared from Russia and ended up in Germany. It is a spy thriller with well researched background. The story remains true to claims surrounding the disappearance of the crates and provides an interesting twist to this fascinating story.

The Amber Room was so beloved by the Soviet people that they determined in 1979 that there would be an Amber Room again. If they couldn't have the original, they would create a duplicate. Beginning with old black and white photographs of the original room, they began to study how it could be re-created. Everything had to be learned from scratch, because there were no amber masters who understood the required construction processes.

Minute pictures were etched into amber medallions, the correct bees wax had to be determined to hold amber plates together, and decisions were made concerning the shades of dye to be used on the amber. The actual process of putting together such a project and bringing it to completion was undertaken. Determining how to finance the project was decided and the funds were secured. This was a monumental undertaking.

Work began in the early 1980s, but after the fall of the Soviet government, there was no money to complete the project. After spending an estimated $8 million, leaders had

An artist's rendering of a panel from the Amber Room hints at the work that went into creating each panel. The pictures were formed by selecting the exact pieces of amber to fit into a mosaic. As thousands of pieces were placed into the design, the panels came to life. Gold leaf was placed behind some and others were dyed to meet the needs of the artists. Wax was used to hold the pieces together.

more pressing issues and the project was halted. Eventually a German company, Ruhrgas, came up with $3.5 million to finish the project and it was completed for the 300th birthday of St. Petersburg, Russia. (www.sfgate.com/cgi-bin/article.cgi?file=/c/a/2003/05/31/MN139859.DTL&type=prin)

St. Petersburg has always been a city of excesses, but the Amber Room is considered over the top. A museum official, Yuri Dumanshin, looked at its baroque extravagance and called it the world's biggest jewelry box. The museum was opened at 4:00 P. M. on Saturday, May 31, 2003, to the amazement of honored guests. The room glowed in the hues and tones of a sunset. Four elaborate mosaic pictures were made of semi-precious stones. Amber scenes featuring amber people, amber roses, and amber landscapes decorate the walls. 500,000 amber tiles, which had been fitted together like a gigantic jigsaw puzzle, are unbelievable. The room has 1080 square feet of space and amber panels cover three of the 26-foot high walls. Windows cover most of the fourth wall, bringing natural light into the room. The sight is amazing. To many visitors, The Amber Room is simply overwhelming. Yet it is a fitting tribute to a stone that has transcended the ages, changed civilization, and brought great treasures into the world.

Dr. Kathleen Griffin, Associate Provost of St. Petersburg College, Clearwater, Florida, visited the Amber Room in St. Petersburg, Russia, in 2003, to celebrate St. Petersburg's 300th birthday and the opening of the newly recreated Amber Room.

One of the largest works of jeweled art gleams in the light. Visitors to the Amber Room are still amazed at its beauty and intricate workmanship. Amber panels over twenty feet high cover three magnificent walls.

The details in the Amber Room stagger the visitor who is surprised that such beauty could have been recreated.

The ceilings in the Amber Room are so high that the top and bottom of the walls are difficult to get into the same photo. Looking up is worth the effort as the ceilings are equally beautiful. Most visitors say that "the room" goes over the top.

Jurassic Park®

The love of amber was rekindled when a book, and later a movie, *Jurassic Park®*, were released. The book was published in 1990 and the movie followed in 1993. The story centered on the creation of a Disneyland-type park, Jurassic Park, which contained living dinosaurs.

The dinosaurs in the movie were the creation of John Hammond, the founder of International Genetic Technologies, which created and developed a zoological preserve on an island 120 miles off the west coast of Costa Rica. His company was successful in extracting blood from mosquitoes trapped in amber during the Jurassic Period. The blood came from extinct dinosaurs but still had viable DNA. The DNA was used to create living animals that were featured in the movie.

The movie, a thriller with unexpected turns and tragedies, fascinated audiences. It made amber, particularly amber containing insects, an overnight sensation. What mystery lurks in the jewelry in your jewelry box?

As with many successful movies, there was a sequel. The love of amber as well as scientific interest in amber was intensified. *Jurassic Park®* now is considered a classic movie and still has fans in the reading, viewing and scientific communities.

More than 1000 extinct species of insects have been identified in amber. Baltic amber has been the preferred amber by jewelry enthusiasts, but since Dominican amber has more inclusions, its popularity has grown.

Microbiologist Raul Cano announced in 1995 that he had revived bacterial spores from a sting less bee that had been preserved in amber for 20 to 40 million years. Since the amber completely encased the bee, it protected the bee and the bacteria from destruction. This bacterium was similar to living bacteria, Bacillus sphareicus, which in difficult times ceases moving, eating and reproducing thus becoming dormant without the need for air, water or nourishment. His work has been questioned by other scientists but it is still ongoing at California's Polytechnic State University, San Louis Obispo, California. George O. Poinar Jr. at the University of California, Berkley found that amber preserves the plant or animal tissue, even the stomach contents of the animals. (Rice, 1993, p. vii) In June 1993, Nature Journal reported the recovery of DNA from a weevil, now extinct, that lived over 120 million years ago. (Rice1993, p. ii) There have been many other studies undertaken that one day could lead to a real Jurassic Park ®complete with the possibility for tragedy. (www.emporia.edu/earthsci/amber/myths.htm)

Johnathan has been fascinated by dinosaurs for several years. The movie *Jurassic Park®* has been his favorite. Because of the importance amber played in re-creating dinosaurs in the movie, he is equally fascinated by amber.

Bibliography

Books

Baker, Lillian, (1995), *100 Years of Collectible Jewelry: 1850-1950*, Paducah, KY: Collector Books a Division of Schroeder Publishing co., Inc.

Baker, Lillian, (1997) *Art Nouveau & Art Deco Jewelry: An Identification and Value Guide*, Paducah, KY: Collector Books a Division of Schroeder Publishing co., Inc.

Bauer, Max, (1968) *Precious Stones Volume II*, New York, NY: Dover Publications.

Berry, Steve, (2006) *The Amber Room*, New York, NY: Ballantine Books published by the Random House Group.

Conway, D.J., (1999) *Crystal Enchantments: A Complete Guide to Stones and Their Magical Properties*, Freedom, California: The Crossing Press.

Curzo, Cipriani and Alessandro Borelli (1986) *Simon & Schusters Guide to Gems and Precious Stones*, New York, NY: A Fireside Book published by Simon & Schuster.

Elsbeth, Marguerite (2002), *Crystal Medicine*, St. Paul, Minnesota: Llewellyn Publications.

Grimaldi, ET AL: "Burmese Amber," *American Museum Novitates* No.3361, pp 1-5.

Mottana, Annibale, Rodolfo Crespi, and Giuseppe Liborio, (1977) *Guide to Rocks and Minerals*, New York, NY: A Fireside Book published by Simon and Schuster.

Parkinson, Cornelia, (1988) *Gem Magic: The Wonder of Your Birthstone*, New York, NY: A Fawcett Columbine Book published by Ballantine Books.

Phillips, Clare, (1996) *Jewelry from Antiquity to the Present*, New York, NY: Thames and Hudson.

Rice, Patty, (1993) Amber, The Golden Gem of the Ages, New York: The Kosciuszko Foundation, Inc.

Sinkankas, John, (1966) *Gemstones of North America*, Princeton, NJ: D. Van Nostrand Company, Inc.

Schumann, Walter, (1993) *Handbook of Rocks, Minerals, and Gemstones*, Boston, MA and NY: Houghton Mifflin Co.

Schumann, Walter, (1997) *Gemstones of the World, Revised and Expanded Edition*, New York, NY: Sterling Publishing Co., Inc.

Simon & Schuster, *Guide to Gems*, p. 308.

Simmons, Robert and Naisha Ahsian, (2005) *The Book of Stones: Who They Are and What They Teach*, East Montpelier, VT: Heaven and Earth Publishing LLC.

Williams, Mabel and Marcia Dalphin, editors (1938), The Junior Classics, Volume 3, Myths and Legends, *Phaeton By Thomas Bulfinch*, U.S.: P.F. Collier and Son Corporation.

Websites

www.amberjewelry.com/SearchResults.asp?Cat=89, "Amber Myths and History of Amber", retrieved September 29, 2007

www.amberlady.com/article.htm, retrieved May 2007

www.ambervarnish.com/index.php?pr=secrets, "Secrets of the Old Masters", retrieved May 17, 2008

www.chiapasambercreations.com, retrieved May 2007

www.crlc.ca/crlcart5.htm, Calgary Rock and Lapidary Club, Lapidary journal, "Amber" by J.P. Jutras, retrieved February 1, 2008

www.emporia.edu/earthsci/amber, retrieved May 2007

www.en.wikipedia.org/wiki/Amber_Road, retrieved May 2007

www.foxnews.com/story/0,2933,349028,0.html, "High-Powered X-Rays Reveal Bugs Hidden in Amber", by Clara Moskowitz, Thursday, April 10, 2008, retrieved April 12, 2008

www.ganoskin.com/orchid/archive/199702/msgooo73.htm, retrieved May 2007

www.gdansk.pl/en/article.php?category=453&article=942 &history, retrieved May 2007

www.gemsociety.org/priceg.htm, retrieved May 2007

www.gemstone.org/gem-by-gem/english/amber.htm, retrieved May 2007

www.gplatt.demon.co.uk/abrief.htm,"A Brief human History of Amber, retrieved September/29/2007

www.jcjonline.com/index.asp?layout=articlePrint&article ID=CA6452101, JCKonline.com, "Dateline Gdansk, Not your Grandmother's Amber Jewelry", retrieved January 18, 2008

www.jamescgroves.com/germanambervarnish.htm, "16th Century Amber Varnish and Venetian Amber Varnish – Amber Varnishes available during the 1500 – 1600's", retrieved May 17, 2008

www.khulsey.com/jewelry/kh_jewelry_amber_mining.html, "All About Gemstones: Polish Amber/ Amber Mining and Fishing", retrieved June 12, 2008

www.lapidaryjournal.com/jj/899jj.cfm, "Carving Amber", by Yoli Rose, retrieved February 1, 2008

www.madehow.com/Volume-7/Amber.html, "How Products are Made, Amber Background", retrieved October 4, 2007

www.sfgate.com/cgi-bin/article.cgi?file=/c/a/2003/05/31/ MN139859.DTL&type=prin, retrieved May 2007

www.urban4est.net/storAmberJewelery.htm, retrieved May 2007

www.waddingtons.ca/pages/home/index.php?c=feat_ref/ feat_08_99.php,"Jewellery by Georg Jensen", retrieved June 12, 2008

Index